why fish fart

***and* Other Useless or Gross Information About the World**

Also by Francesca Gould

Why You Shouldn't Eat Your Boogers and Other Useless or Gross Information About Your Body

why fish fart

and Other Useless or Gross Information About the World

francesca gould

JEREMY P. TARCHER/PENGUIN
a member of Penguin Group (USA) Inc.
New York

JEREMY P. TARCHER/PENGUIN
Published by the Penguin Group
Penguin Group (USA) Inc., 375 Hudson Street, New York, New York
10014, USA • Penguin Group (Canada), 90 Eglinton Avenue East, Suite 700,
Toronto, Ontario M4P 2Y3, Canada (a division of Pearson Canada Inc.) • Penguin
Books Ltd, 80 Strand, London WC2R 0RL, England • Penguin Ireland,
25 St Stephen's Green, Dublin 2, Ireland (a division of Penguin Books Ltd) • Penguin
Group (Australia), 250 Camberwell Road, Camberwell, Victoria 3124, Australia
(a division of Pearson Australia Group Pty Ltd) • Penguin Books India Pvt Ltd,
11 Community Centre, Panchsheel Park, New Delhi–110 017, India • Penguin
Group (NZ), 67 Apollo Drive, Rosedale, North Shore 0632, New Zealand (a division of
Pearson New Zealand Ltd) • Penguin Books (South Africa) (Pty) Ltd,
24 Sturdee Avenue, Rosebank, Johannesburg 2196, South Africa

Penguin Books Ltd, Registered Offices: 80 Strand, London WC2R 0RL, England

Library of Congress Cataloging-in-Publication Data

Gould, Francesca.
Why fish fart and other useless or gross information
about the world / Francesca Gould.
p. cm.
ISBN 978-1-58542-757-4
1. Curiosities and wonders. I. Title.
AG243.G63 2009 2009023253
031.02—dc22

Printed in the United States of America
1 3 5 7 9 10 8 6 4 2

BOOK DESIGN BY NICOLE LAROCHE

This book is dedicated to Andrew Bedale
and my gorgeous daughter Sienna

The author would like to thank David Haviland for his help in the production of the book; also Dr. Richard Pollard, a highly amusing entomologist who gave invaluable help regarding the eating of insects and also advised reading up on fig wasps while snacking on Fig Newtons.

contents

why fish fart

and Other Useless or Gross
Information About the World

chapter one

obscene cuisine

how do you make "bird's nest soup"?

The Chinese delicacy "bird's nest soup" is one of those rare dishes that sounds incredibly gross but is in fact far, far worse. This dish has been highly esteemed since the Ming dynasty, and it is said there is no higher honor one can bestow upon a guest than to serve them bird's nest soup. The soup is prized for its rich nutrient content and purported health benefits. This rare and expensive soup is made from the nests of a certain type of swift that lives in caves. The nests themselves are made from the bird's saliva, and are formed high up in caves around Southeast Asia. This saliva sets into a solid, rubbery substance that helps to glue the twigs to the cave roofs.

Collecting the swiftlets' nests from the bat-filled caves is a very ancient and very dangerous job. Nest collectors have to climb extremely high, and use long bamboo poles to remove the nests, which are found stuck to the roofs of the caves.

To make bird's nest soup, chefs simmer the nests in chicken broth for hours, until they become rubbery. Once cooked, the nests are said to be chewy and fairly tasteless, which is why flavorings such as chicken stock are often added. One reason why this culinary delight is so highly prized is that it is believed to be an aphrodisiac, as well as being beneficial for lung problems. For centuries, the Chinese have encouraged their children to eat the soup, believing it will help them to grow.

who eats tarantula omelettes?

Despite its frightening appearance, the tarantula spider is regarded as a delicacy by a number of cultures around the world. For example, roasted tarantula is eaten by the Bushmen of central Africa, while people in northern Thailand reportedly like to strip off the spiders' legs and roast the bodies. The Piaroa Indians of Venezuela enjoy eating the big, hairy, bird-eating "goliath tarantula" (*Theraphosa blondi*), which has a leg span of 10 inches (25 cm), and an abdomen the size of a tennis ball. The whole thing is the size of a dinner plate!

The Piaroa hunt for tarantulas and when they've caught one, they bend its legs backward over its body and tie them together, so it can be taken back to camp. To prepare the tarantula, a leaf is used to twist off the abdomen (to avoid touching the hairs, which can irritate the skin) and the spider is then rolled in a leaf and roasted in hot coals. Once it's cooked, the spider is eaten by picking out the bits of flesh, rather like eating a crab. Apparently, tarantulas taste a bit like prawns. Bits of the meat can get stuck between your teeth, but luckily the tarantula's long fangs make excellent toothpicks! If the Piaroa manage to catch a female tarantula, they will squeeze the eggs out from her body, wrap them in a leaf, and roast them over a fire to make a tarantula omelette.

The eating of spiders is also very popular in Cambodia and Laos, where they are commonly toasted on bamboo skewers over a fire and served whole with salt or chilies. Alternatively, some people prefer their spiders fried in butter with a clove of garlic.

what is "dancing-eating?"

"Dancing-eating" or, to give it its proper name, "odorigui," is the Japanese practice of eating live animals. One common odorigui dish is a small, transparent fish called "shirouo," which is served in alcohol and washed down with sake. More adventurous diners prefer to feast themselves on live octopus. To prepare this dish, the chef will remove the live octopus from a large tank, slice off one tentacle, and then simply serve it on a plate with some soy sauce. Apparently, the limb continues to writhe and twist on your plate, and when you eat it, the suckers attach themselves to the roof of your mouth.

The Japanese are not alone in their taste for very, very fresh meat. In China, there is a popular dish called "drunken shrimp," which is also eaten live. When the dish arrives, it comes swimming in a bowl of sweet alcohol, which is supposed to help make the shrimp a little less feisty and, because it's effectively in a drunken stupor, prevents it from escaping. The shrimp should be left to swim in the alcohol for about five minutes before being eaten. The normal way to eat drunken shrimp is then to remove it from the bowl using chopsticks and put it on a plate. The diner then removes the shrimp's head with his or her fingers before munching on its twitching body. The Japanese also enjoy a similar dish called "drunken crab."

However, both of these practices sound quite bland when compared with the extraordinary dish served at Pingxiang, on China's border with Vietnam. Diners here buy live monkeys at the market and then take them to a local inn to have them prepared by a chef. This preparation consists of forcing a monkey to drink large amounts of rice wine until it passes out. Then the chefs bind the monkey's limbs, chop open its skull, and scoop out the brains into a bowl. Apparently, the test of a well-prepared monkey brain is that the blood vessels should still be pulsing when the dish is served. The brains are eaten with condiments, including pickled ginger, chili pepper, fried peanuts, and coriander. It is said that monkey brain tastes like tofu, which rather begs the question—wouldn't it be simpler to just buy some tofu?

which dish, properly prepared, should contain just enough poison to numb your lips?

The answer to this is the Japanese dish "fugu," which is also known as "puffer fish" or "blowfish." Fugu is regarded as an exceptional delicacy in Japan, and many liken the taste to chicken, although it is eaten more for the thrill than the taste, as fugu contains one of the most powerful poisons found in nature. The fish's traditional nickname is "teppo," which means pistol.

In its natural environment, fugu is a peculiar-looking fish that can puff itself up into a large, round ball when threatened by predators. Beyond this, its main form of defense is the lethal poison called "tetrodotoxin," which is contained in its internal organs. One blowfish contains enough tetrodotoxin to kill thirty adult humans.

Consequently, fugu chefs need to be exceptionally skilled and precise, to ensure that all the poisonous parts of the fish are removed before serving. In Japan, fugu restaurants specialize in its preparation and the removal of the fish's deadly poison. If the fish has not been prepared correctly, with all of its poison removed, the toxins can destroy the nerve tissue inside a person's body, paralyzing the muscles necessary to breathe and causing death within about four to six hours. The most poisonous part of the puffer fish is the liver, which should be completely removed during preparation. Apparently, a skilled fugu chef will leave just enough poison to numb your lips. Even so, it's probably best not to annoy the chef.

why is civet poo coffee so expensive?

The world's most expensive coffee is an Indonesian specialty called "Kopi Luwak," which is believed to be the best in the world and is very popular in Japan and America. Bizarrely, this coffee is made from beans that have been eaten and then excreted by the common palm civet (*Paradoxurus hermaphroditus*).

The palm civet is a small, bobcat-like mammal that lives in certain rainforest regions of Indonesia. The palm civet's main diet consists of berries, insects, and other small mammals, but it also enjoys eating coffee beans. However, not just any old coffee beans will do; the fussy civet chooses only the reddest and ripest beans. These beans then pass almost unchanged through the palm civet's digestive system before being pooed out and then harvested by unfortunate plantation workers before being sold for ludicrous sums to wealthy coffee fanatics. To buy just 2 ounces (57 g) of Kopi Luwak coffee, which is enough to make only one cup, costs around $37.00 (£24.00), making it a very expensive way to start your day!

But what is it that makes Kopi Luwak taste so good? There seem to be two main possibilities. The exceptional flavor may result from the effect of the civet's digestive juices on the coffee beans. Alternatively, these digestive juices may have no real effect, and the answer may simply lie in the civet's fastidious pickiness in selecting only the very ripest, reddest coffee beans.

which is the world's tastiest insect?

For centuries, many cultures around the world have been eating all sorts of insects and bugs. The practice even has a posh name—"entomophagy." In some respects, insects make quite a good meal. They are widely available, easy to cook, and highly nutritious; caterpillars, for example, are chock-full of protein and iron. Bugs that are commonly munched around the world include flies, beetles, dragonflies, grasshoppers, cockroaches, butterflies, moths, mosquitoes, ants, and even bees and wasps. In fact, there are more than 1,400 species of edible insects.

In Nigeria, people enjoy snacking on crickets that are disemboweled and then roasted over open coals. Grasshoppers are also popular, and people working the fields will often eat them raw. In China, a nice warm bowl of earthworm soup is believed to help treat fever. On the coast of Macao, you can buy a bag of fried cockroaches. And on the streets of Bangkok, in Thailand, fried grasshoppers are sold in much the same way.

In fact, people in Thailand eat a wide variety of insects, including cicadas, locusts, mantises, deep-fried crickets, grasshoppers (usually fried), dung beetles, moth and butterfly pupae, wasp and bee larvae, termites (in soup), giant water bugs (steamed), bamboo borers (these are small grubs that are sometimes described on menus as "fried little white babies"), weaver ants (which are eaten raw or rubbed with salt, chili, or pepper) and their eggs (these can also be eaten as a paste), and even the revolting-sounding grilled tarantula.

Meanwhile, in South America, tree worms and various stingless bees and wasps are widely eaten in Brazil. They are apparently known for their pleasant, almondy taste. In Columbia, culonas ants are also enjoyed for their nutty flavor. These winged ants are collected with large sweeping brooms, placed into boiling water, then taken out and dried on the grill. And in the United States, worker ants of this species are often dipped into chocolate.

One company in California makes products such as the "Cricket Lick-It," which is a tasty, sugar-free lollipop with, you've guessed it, a real cricket in the center. Perhaps you'd prefer the "Scorpion Sucker," which comes in an array of colors and flavors—such as blueberry and banana—and contains, predictably enough, a scorpion. Products from other companies include the "Antlix," a peppermint-flavored lollipop, that contains real farm-raised ants—apparently, they taste similar to chili peppers, and are said to be good for giving you an energy boost; the "Tequilalix," a tequila-flavored lollipop, which contains a real worm; and the "Vodkalix," a vodka-flavored lollipop, which also for good measure contains a scorpion. The scorpions are said to be heat-treated to remove any toxins, to make sure they are safe to eat. If you're not feeling that adventurous, how about a tasty snack of worms? These come in a variety of flavors, including barbecue, cheddar cheese, and Mexican spice.

However, if I had to choose which insect would be the most delicious, I think I might plump for the Australian honey ant. This insect is greatly prized by the Aborigines, who call them by the rather appetizing name "yarrumpa." This insect stores so much of a sugary fluid in its body that its hind end swells up into a ball that is big enough to eat. People bite the bug's end off to savor the sweet stuff found inside. They say it's just like eating honey, only crunchier.

why did nelson's navy eat their biscuits in the dark?

In the 1700s, under Horatio Nelson's leadership, Britain had the most powerful navy in the world. However, conditions on board were truly disgusting and navy discipline was harsh. Floggings were frequent, the pay was low, hygiene was poor, and the food was often infested with bugs.

Before setting off, ships would be loaded with food, including fruit, vegetables, and live animals, but once the ship departed, it wouldn't be long before all the animals had been slaughtered and the remaining food had rotted. As bread rapidly became moldy, the sailors instead ate "ship's biscuits," also known as "hardtack," which were made with flour, water, and salt. These biscuits would often contain weevils or maggots, so the sailors would tap the biscuits on the table to try to knock out most of the bugs before eating. For this reason, many sailors would wait until dark to eat them, so they wouldn't have to see the maggots crawling inside.

The sailors' diet also included salted meat that was so hard that it was practically inedible, even after being soaked and then boiled for hours. The meat was so dark and hard that some sailors developed a weird new craft—they would carve objects out of the meat and then polish their sculptures. Cheese was preserved by dipping it in tar, which made it taste disgusting. One of the main constituents of a sailor's diet was a healthy ration of a gallon of beer a day. Although beer did not keep aboard ship for very long, it was drunk as a safer alternative to water, as water on board the ship would quickly become green and slimy. A ten-year-old boy who served at Trafalgar wrote this letter home about the food:

> We live on beef which has been ten to eleven years in the cask and on biscuit which makes your throat cold in eating owing to the maggots which are very cold when you eat them, like calvesfoot jelly.

However, there was of course some fresh food such as fish, dolphins, sharks, and birds available from the surrounding waters. Tortoises and turtles were also highly prized, as they could be kept alive for many weeks in the hold, without food or water, providing another source of fresh meat for the later stages of a voyage.

And God forbid that you should fall ill. According to Tobias Smollett, a ship's doctor during this time, the sick berth on board was more likely to kill than cure. He wrote that patients were kept far below deck, deprived of daylight and fresh air, and breathing "nothing but the morbid steams exhaling from their own excrement and diseased bodies, devoured with vermin hatched in the filth that surround them."

what is the "king of fruit"?

The durian is a peculiar-looking, disgusting-smelling, melon-like fruit that is nonetheless considered a delicacy by millions of people in the Far East. In Indonesia, Malaysia, and especially Thailand, it is affectionately known as the "King of Fruit" and, despite its disgusting smell, it is widely sold in Southeast Asian markets.

The durian fruit can weigh as much as 10 pounds (4.5 kg), and has a hard green shell studded with spikes. The fruit grows at the top of durian trees, which can reach up to 164 feet (50 m) in height. As if this fruit were not unpleasant enough in its appearance and odor, it also presents a physical danger, as it has been known to drop from the branches and kill people below.

Apparently, the durian has a pleasant taste, which is described as being like a rich, buttery custard, highly flavored with almonds. Unfortunately, however, many people find the smell far too off-putting to consider actually tasting the thing. This smell is variously described as being like excrement, stale vomit, or even like a dead rat decomposing in a plate of vomit. Travel and food writer Richard Sterling memorably described the durian as smelling like "pigshit, turpentine and onions, garnished with a gym sock."

Because of the durian fruit's awful smell, the governments of many Asian countries have banned people from taking the fruit onto public transport. It is also banned by most hotels and airlines in the region, and rental car agencies will reportedly fine you if the car stinks of it when you hand in the keys.

However, the world of nature contains numerous other putrid plants that are just as stinky and vile as the durian. There is the flower of the exotic plant *Amorphophallus titanum*, from Sumatra, which is also referred to as the "corpse flower." This impressive plant has large petals and a phallic-like spadix in the center. It can reach a height of 8 feet (2.5 m), but its most notable feature is its nauseating stench, which has been described as a mixture of excrement and rotting flesh. The foul smell is designed to attract flies, which help to pollinate the plant, and can be detected up to 0.5 miles (0.8 km) away.

Aristolochia Grandiflora, also known as the "Pelican Flower," is found in Brazil and the Caribbean.

This alien-looking plant is a huge flower, which can reach more than 3 feet (1 m) in length. However, its smell is so disgusting that wild animals actually stampede from the area when the plant blossoms.

The flower *Rafflesia Arnoldii* is named after Sir Thomas Raffles, who was the founder of Singapore. It is a huge, impressive-looking plant, with five mottled, orange-brown-and-white petals; the flower can measure up to 3 feet (1 m) across. Unfortunately for Raffles, the smell of this flower has been described as being like rotten flesh, and because of this it is popularly known as the "stinking corpse lily."

A pleasant amble through the woodlands of Europe or North America can quickly become disrupted if you come across a group of stinkhorns. These disgusting-smelling fungi grow in rotting wood and produce a smell that is said to be like decaying flesh. Their thick stalks, which grow to a height of about 10 inches (25 cm), are capped with a liquid ball of a jellylike, foul-smelling green or brown slime. Flies and other insects are attracted by the smell and wallow in the slime, becoming covered with fungal spores. They then carry the spores when they fly away, and in this way help the stinkhorns to disperse.

how do you make a fly burger?

Lake Victoria in Africa is the second-biggest freshwater lake in the world. At certain times of the year, peculiar dense clouds appear and hover over the lake. These clouds are made up of trillions of lake flies, which make up the largest swarm on Earth. These swarming gnats are called "E sami" by the local people, and the clouds are visible on the horizon from miles away. Unfortunately, increased pollution and sewage are helping to increase their numbers.

Many of the flies reach land, and when this happens, the air becomes thick with flies, but the locals still carry on with their normal, daily lives. In fact, the midges are positively welcomed, as people take advantage of this blight by catching the flies for food. Entire villages will come together to catch the flies in items such as nets, baskets, pots, and frying pans. One popular way of catching the flies is simply to dampen the inside of a pan—the flies stick to it, and can then be cooked.

Each village has its own recipe, but one of the most popular and widespread is the fly burger, which is also extremely nutritious. Fly burgers are made by crushing a handful of flies, which are then molded into the shape of a burger and left to dry in the sun. The burger is then cooked—some regions like to roast them, while others enjoy their fly burgers deep-fried. Each burger contains about half a million flies, and seven times more protein than a hamburger.

what is a cod worm?

Have you ever been tucking into your fish and chips and noticed a worm inside your fish? If so, it was probably a type of round worm called a cod worm (*Phocanema decipiens*). Cod worms can grow to about 1.5 inches (4 cm) long, and vary in color from creamy white to dark brown. Although they are unsightly, they are actually quite harmless.

The life cycle of the cod worm is a fascinating and complex one. The adult cod worms live and mate inside the stomachs of gray seals, specifically gray seals that have eaten worm-infested fish. The cod worm's eggs pass out of the seal and into the ocean via the seal's feces. The eggs then hatch into tiny worm larvae that get eaten by small crustaceans such as shrimp. In turn, these small crustaceans get eaten by larger sea creatures, including fish such as cod and haddock. As the cod worm larvae reach the fish's stomach, they continue to grow, and eventually burrow through the stomach wall and into the flesh of the fish. Finally, the cod worm's life cycle comes full circle when the fish is swallowed by another seal, and the process begins again.

Although it must be quite unpleasant to find a worm in your dinner, cod worms are relatively harmless. The worms are killed by cooking or freezing the fish, and there is no evidence that anyone has ever had an illness associated with the cod worm. Occasionally, when you buy fresh fish, you might even find a live cod worm inside it; but even so, although they're pretty gross, they are harmless if eaten. And the incidence of infected fish is very small in relation to the thousands of tons of fish landed each year.

what are "sweetbreads"?

"Sweetbreads" is a culinary term that refers to an animal's internal organs. The most commonly served sweetbreads are the thymus gland and pancreas, but chefs are also known to prepare the salivary and lymphatic glands. The thymus gland, which is known as the "throat sweetbreads," is found near the base of an animal's neck. The pancreas, which we call the "heart sweetbreads," is attached to the last rib and lies near the heart. Generally, the throat sweetbreads of younger animals are preferred because of their firm texture and delicate flavor. Sweetbreads can also be derived from lamb, calf, and pigs.

So why are they called sweetbreads? The most likely answer seems to be simply that chefs realized that "thymus" and "pancreas" were not particularly appetizing-sounding words on a menu, so instead they use the rather vague, pleasant-sounding term "sweetbreads."

what goes into a sausage?

After a pig has been slaughtered and the various sections of meat, such as hams and bacon, have been taken away and processed, the remains of the carcass have other uses. The intestines are used as casings for sausages and hot dogs. The trotters are sold fresh, and the pig's bones are used to make gelatine, which is used to thicken foods such as yogurts and sweets. The remaining edible parts of the pig, including snout, ears, face (including eyelids and lips), heart, kidneys, liver, stomach, and tail, all end up in sausages and hot dogs.

> *Laws are like sausages. It's better not to see them being made.*
>
> OTTO VON BISMARCK,
> *German politician (1815–98)*

which dish contains six types of penis?

Most of us delicate Westerners would probably consider eating the genitals of an animal to be rather gross (unless it was part of a sausage, obviously), but not the Chinese. Some of these fearless gourmands choose to feast on animal penis, either steamed, roasted, or boiled. Fans of this unusual branch of gastronomy say that penises have numerous health benefits—they are low in cholesterol, they boost a man's sex drive, and they are also helpful for treating all sorts of ailments—although these claims do not seem to have been fully embraced by the scientific community.

So where, you're no doubt wondering, can I sample this delightful Chinese cuisine? At China's first specialty penis restaurant, of course. As consuming a penis is supposed to help boost virility, you will find a predominantly male clientele at this establishment. It is common to see businessmen and government officials dining here, and these lucky chaps are able to sample the members of dogs, deer, yaks, oxen, donkeys, horses, and even seals. One popular dish consists of a goat's penis, which is sliced, dipped in flour, fried, and served skewered with soy sauce. Another favorite is the hotpot, which contains a range of samples of what the restaurant has to offer: six types of penis and four types of testicle, all simmered expertly in chicken stock.

which cocktail contains a human toe?

In Canada, 65,000 hardy souls have joined the elite membership of the Sourtoe Cocktail Club, by drinking a very unusual cocktail, into which the barman places a preserved human toe. According to the rules of the club, to become a true "Sourtoer," the drinker must allow the toe to touch his or her lips while drinking. Pass this test and you will receive a certificate to mark your achievement, as well as a story guaranteed to impress, or perhaps repulse, your friends and family for years to come.

So how did this bizarre practice start? And, more to the point, whose is the toe? According to the Sourtoe Cocktail Club (www.sourtoecocktailclub.com), there has been a succession of toes donated for this purpose. . . .

The first toe belonged to trapper Louie Liken, for a time at least. In the 1920s, Liken became stuck in a blizzard with his brother Otto, and his big toe froze. To prevent the onset of gangrene, the toe would have to be amputated. However, Louie didn't trust doctors, and definitely didn't want to have to pay one, and most certainly was not prepared to make the 60-mile (96.5-km) journey to Dawson to find one. Instead, he drank as much rum as he could manage, to anesthetize the pain, and then got his brother to chop off the toe with his ax.

The toe was discovered many years later by Captain Dick Stevenson while cleaning the brothers' cabin, and it was he who ingeniously came up with the idea of the "Sourtoe Cocktail," and started serving it at a hotel in Canada in 1973. However, in July 1980, a man called Garry Younger was trying for the Sourtoe record, and on his thirteenth glass of Sourtoe Cocktail, his chair tipped over backward and he swallowed the toe. Of course, it was never recovered, so another one was needed.

Toe number two was donated by a lady named Mrs. Lawrence, whose middle toe had been amputated many years before. The toe was kept in a jar of salt in the bar but sadly got lost during a renovation.

Toe number three was donated by another trapper, who had had to have it removed due to frostbite. Unfortunately, in 1983, the toe was swallowed by a baseball player from the Northwest Territories. But not to worry because . . .

Toe number four was sent in by an anonymous donor but was later stolen. The thief was identified but refused to give the toe back. Only when the police threatened to charge him with the transportation of human body parts across the border did he agree to return the pilfered digit.

Toes five and six were donated by another anonymous donor with medical connections, who gave the toes on the condition that his nurses got to drink the Sourtoe for free. The fate of these toes is uncertain.

Toe number seven had been amputated from a diabetes sufferer who decided to help out after reading about the Sourtoe plight in the newspaper.

And finally, the most recent toe, toe number eight, was donated by a person who, as a result of a lawnmower accident, had his toe cut off while wearing open-toed sandals.

do flies really puke on your food?

The *Musca domestica*, better known as the common housefly, is found in almost every part of the world. It will eat almost anything, including rotting vegetables, animal carcasses, excrement, and vomit. This hairy six-legged insect's mouthparts include a proboscis, which is specially adapted for sucking up fluid or semifluid foods. The proboscis ends in a pair of oval-shaped, fleshy labella through which it sucks up the food.

However, the housefly can suck up only foods that are in a somewhat-liquid state. This means the fly has no problem sucking up thin fluids, such as milk or beer, or runny solids, such as poo, spit, or mucus—the fly simply places its labella onto the food and sucks it up. However, when the housefly's dinner consists of something rather more solid, such as dried blood, cheese, or cooked meats, it can't just suck these up. Instead, it has to moisten these foods, using either spit or the regurgitated contents of a previous meal—perhaps some partially digested rotten rubbish or dog poo, for instance. As well as providing moisture, this fly vomit contains an acid that helps to dissolve whatever the fly has landed on, making it possible for the fly to hoover it up.

The housefly has three pairs of legs, which end in claws, and a pair of fleshy pad-like structures called the pulvilli, which contain tiny hairs. These sticky hairs enable the fly to stick to even very smooth surfaces, such as glass. The pulvilli are also responsible for picking up harmful germs when the fly lands on things like dog poo, mucus, and rubbish. As a result, flies can carry more than a hundred types of disease-causing bacteria, including those that cause diarrhea, cholera, typhoid, and food poisoning.

what is maggot cheese?

Many of us enjoy cheeses that are riddled with mold, such as Stilton, but we would probably draw the line at a cheese infested with squirming maggots. But not the residents of Sardinia . . .

"Casu marzu," which literally means rotten cheese, is a Sardinian tradition that more than lives up to its name. It is traditionally made with sheep's milk, into which flies have deliberately been allowed to lay thousands of eggs. When these eggs hatch, they produce translucent white worms, which grow up to around 8 millimeters long. These worms produce a substance that causes the fat in the cheese to putrefy. The brown, decomposing cheese becomes a soft, sticky mass that creates a spicy, burning sensation in the mouth when eaten.

Eating insect-ridden food presents a range of new challenges. For one thing, the maggot larvae can leap for distances of up to 6 inches (15 cm) and have been known to jump into the cheese eater's eyes. For this reason, casu marzu eaters often first seal the cheese in a bag. This starves the maggots of oxygen, causing them to leap out from the cheese and squirm at the bottom of the bag.

Because of the obvious health risks, casu marzu is officially outlawed in Italy, but it can still be bought on the black market. One official from the Sardinian health department stated that anyone caught selling the cheese would be heavily fined, but admitted, "As a Sardinian and a man, let me tell you, I have never heard of anyone falling ill after eating this stuff. Sometimes, it tastes real good."

Sometimes extraordinary measures are taken to improve the flavor of cheese. In 1888, author J. G. Bourke received a letter from Dr. Gustav Jaeger, which informed him about an unusual cheese that had been sold in Berlin. Apparently, a storekeeper was punished when it was revealed that he had been using the urine of young girls to make his cheese richer and more tasty. It was said that the cheese had proved very popular and people had loved the taste.

what are "one-hundred-year-old eggs"?

China's cuisine is not for the fainthearted. From the skinned dogs displayed at markets to the scorpion kebabs sold by traders, the Chinese display an admirably open-minded approach to food, which puts our prissy Western palates to shame. So it's no surprise to find out they also enjoy eating a dish called "pi-tan," which is also known as "one-hundred-year-old eggs." Despite the rather attention-seeking name, pi-tan eggs are actually only four years old and are considered a great delicacy.

To make pi-tan, the Chinese usually use duck eggs that are left to soak in a mixture containing salt, lime, and tea leaves for three months. After this, the eggs are then coated with a paste of clay, lime, ashes, and salt, and buried in the ground. They are left there for three or four years. After this time, the eggs are dug up and then peeled to reveal a green, cheesy yolk and a yellow, gelatinous egg white. The raw, foul-smelling eggs are then dipped in vinegar and eaten.

what is "tapping the admiral"?

In 1805, Admiral Horatio Nelson died in the Battle of Trafalgar on his ship, *Victory*, and his body was shipped home to England for burial. There is a common mistaken belief that Nelson's body was transported back to England in a keg of rum. For that reason, to this day, rum is colloquially known as "Nelson's Blood."

However, it was in fact brandy, not rum, that preserved Nelson's body on the long journey back to England. At this time, most sailors who died at sea would also be buried at sea, so preserving Nelson's body presented an unusual challenge. It was the idea of the ship's doctor, Dr. Beatty, to preserve Nelson's body by storing it in a barrel of brandy. According to legend, when the barrel was opened on the ship's return to England, most of the brandy was gone. Apparently, the sailors on board had been tapping the barrel for drinks of brandy, and this is where the saying "tapping the Admiral"—which means illicit drinking—comes from.

chapter two

weird creatures

why does the "horned lizard" squirt blood from its eyes?

In 1960, American cinema audiences were scared witless by the astonishingly realistic dinosaurs in the film *The Lost World.* In fact, what appeared to be enormous, snarling predators were actually "horned lizards," small lizards just a few inches long, which were filmed using powerful magnification. There are thirteen species of horned lizard found in North America, and they have broad bodies and rough skin, with horns on their head and spines across their back. Although they look quite frightening, horned lizards are usually docile and passive unless provoked, and they have a rather gross way of fending off predators.

Despite their spiky, unappetizing features, horned lizards are preyed upon by a wide range of creatures, including hawks, roadrunners, snakes, dogs, wolves, coyotes, and other lizards. To defend themselves against these numerous threats, the horned lizard has developed a number of remarkable and unusual talents. When faced with a hungry predator, the horned lizard is able to inflate itself with air so that it resembles a spiky balloon, making it look bigger and more threatening.

If this doesn't work, the lizard can employ another

bizarre technique by shooting blood from its eyes. It arches its back defensively and closes its eyes. The blood pressure inside the lizard's head quickly rises, and this pressure causes blood vessels in the sinuses to rupture. Then the lizard's eyeballs become infused with blood and swell up, causing the eyelids to bulge. Finally, the swollen eyelids, engorged with blood, send out a fine spray of blood through the tear ducts. They can shoot blood from either one or both eyes, up to a distance of 6 feet (2 m), and they can even direct the spray forward or backward.

So what's the point of this extraordinary ability? Well, first, the blood squirting serves to alarm and confuse would-be predators, naturally enough. Another benefit is that the stream of blood contains a chemical that is noxious to dogs, wolves, and coyotes. If any of these dogs gets the blood in its mouth, it shakes its head vigorously from side to side, as it finds the substance so repulsive.

However, despite having these spectacular talents, the horned lizards' most effective way of avoiding predators is simply to lie still. The lizards' coloring is similar to that of the soil on which they live, and they can get rid of any shadows by flattening their bodies against the ground. If they have to move, they can run and then stop suddenly and unexpectedly, lying flat to merge into their surroundings, leaving the predator scratching its head, wondering where its dinner went.

which worm may crawl out of your nose?

Worldwide, about 1.4 billion people have got one or more Ascaris worms living inside them. However, they rarely cause a problem, so people don't realize they're playing host to an uninvited guest. The most commonly encountered parasitic worm is the *Ascaris lumbricoides*, or "roundworm," which can measure 6–14 inches (15–35.5 cm) in length, making it about the size of an earthworm. So how do they find their way into our bodies? The eggs of Ascaris are transmitted in feces and find their way into a new host via food or dirty fingers.

Once inside the body, the eggs hatch into larvae within a person's stomach, before moving into the small intestine. They then penetrate through the intestinal wall, enter the blood circulation, and are carried around the body. The larvae make their way to the lungs, and as they continue to climb, the young Ascaris larvae tend to slightly irritate the airways as they crawl up the back of the throat, which can cause a cough reflex, resulting in a free ride into the throat. Occasionally adult worms can be coughed up into the mouth too, which gives the sufferer quite a shock. From the throat, the larvae are swallowed, which brings them to the intestines, where they make their home and grow into adult worms.

Adult worms can live in a person's gut for about six to twenty-four months, and many people are unaware of their infestation until they see an adult worm in their poo or cough up an adult worm. The worms have even also been known to come out of a person's nose or in their vomit. In severe cases, where people host many worms, the victim may suffer from anemia or even malnutrition. The Ascaris worm is estimated to affect 4 million people in the United States.

which fish uses slime to kill its enemies?

When a creature has not one disgusting name but two, the odds are against it being a cute, doe-eyed bundle of fur. Such is the case with the hagfish, also known as the "slime eel," which is every bit as disgusting as these names suggest. The hagfish is an ancient creature, which has been around for 300 million years. It is virtually blind and resembles an oversized, slimy worm. Hagfish are completely covered in mucus, which oozes out of hundreds of slime pores found on the sides of their bodies.

The hagfish hangs around on the bottom of the ocean floor and looks for fish that are either sick or dead. It then climbs inside its prey through the mouth, gill, eye socket, or anus, and devours it from the inside out, using its tooth-covered tongue to scrape its victim to pieces. The hagfish's mouth is very unusual and extremely ugly. The mouth is surrounded by four pairs of tentacles, with a single tooth on the roof of the mouth, and a tongue covered with two rows of strong, pointy teeth.

When faced with a predator, the hagfish uses its glands to produce a substantial amount of slime, which quickly turns the surrounding water thick and gunky. This slime surrounds the predator, protecting the hagfish. The thick slime can even kill the predator by clogging up its gills and suffocating it. However, there is a delicate balance to be struck, as the hagfish must be careful not to produce too much mucus, as it can suffocate itself instead. Hagfish slime is a unique substance made from protein, which is best described as a kind of rubbery snot. When the hagfish thinks that the predator has been seen off and the coast is clear, it will tie itself into knots to rub away the excess slime.

Fishermen consider hagfish a nuisance because they penetrate the bodies of fish caught in the fishermen's nets and eat them from the inside out, leaving nothing but skin and bones. One large fish was found with more than one hundred hagfish munching inside it.

what is the hagfish's even uglier cousin?

A close relative of the hagfish is the lamprey. The lamprey has a slimy, eel-like body and can grow up to 4 feet (1.2 m) in length. It has two fins along its back, and a single nostril in the middle of the head, but its mouth is even more unpleasant. The lamprey's large funnel-like mouth is very round, with rows of razor-sharp teeth and a sharp tongue and lips that also bear many rasping teeth. Unlike the hagfish, the lamprey chooses healthy creatures to feed on, including human beings.

When a lamprey comes across a potential victim, which could be another fish or a swimmer with cold legs, it attacks by attaching itself. The lamprey's sucking lips close upon the skin of the victim, and its sharp teeth scrape open a wound. The lamprey has a special type of saliva that stops the prey's blood from clotting. Then the toothed tongue goes into action, lapping the blood and body fluids of the victim. Once the lamprey has eaten its fill, it lets go. The victim doesn't die from the bite itself, but it bears a wound that may become infected. Where lampreys are common, a single host fish may have several lampreys attached to it at one time. And if a fish suffers too many lamprey attacks, it may become weak and die. Infected lamprey bites cause the deaths of many fish in the ocean and the Great

Lakes. However, human victims are fairly rare, with no recorded fatalities.

which animal is the best actor?

One of nature's most impressive and convincing performances is that given by the North American opossum (*Didelphis virginiana*), a small mammal found throughout the United States. Opossums usually live in woodlands and reach about 20 inches (51 cm) long, and are mostly gray-colored, with a white, pointed face and a tail like a rat's but bigger. They are active at night and will eat almost anything: insects, snails, toads, rodents, and even dead animals.

The main threats faced by opossums are dogs, cats, foxes, and people. When they feel threatened, opossums will hiss and bear their many sharp teeth. If this doesn't work, they may use their considerable acting skills, by playing dead. The opossum will flop down on its side, lying still, with its eyes half closed, its mouth hanging open, and its tongue lolling out of its mouth. If required, it can play dead for several hours; even if it is poked or kicked it will not move a muscle. For added verisimilitude, the opossum may even defecate on itself, as well as releasing a foul-smelling green slime, which smells like rotting flesh, from its anus. Most of the opossum's predators will not eat animals that are already dead, so they leave the stinking opossum alone.

why do skunks stink?

The skunk is a member of the weasel family, which is commonly found in North America. Skunks rarely attack unless they are cornered or defending their young. But if a skunk does feel threatened and is unable to escape, it will put on an aggressive display by growling, fluffing up its fur, shaking its tail, stamping the ground with its front feet, standing on its hind legs, and spitting violently, all of which will hopefully serve to frighten and discourage the potential attacker. However, if those methods don't work and the predator remains a threat, the skunk will lift up its tail and spray.

The chemical skunks spray at their enemies is a sulfur compound called N-bulymercaptan, which burns the attacker's skin and causes stinging in the eyes, making the predator blind for a short while. The skunk sprays this disgusting, noxious chemical in a fanlike pattern from two small openings inside its anus, which act rather like tiny water pistols. The glands that produce N-bulymercaptan hold enough for five or six full-powered sprays. A skunk can aim this spray by twisting the nozzles in all directions, but they seldom spray without being provoked or giving warning. The skunk's spray is extremely accurate and effective at a range of up to 15 feet (4.5 m). As a result, even though skunks have sharp teeth, they rarely use them in defense.

If it gets into a person's eyes, skunk spray is extremely irritating and can cause temporary blindness but no permanent damage.

which wasp lays its eggs inside a caterpillar?

There is a tiny wasp called *Cotesia congregata*, commonly found in the United States, that has a deeply unpleasant, albeit fiendishly clever, way of helping her eggs to hatch, which also ensures the little ones have a ready-made meal waiting for them.

The victim of this unpleasant practice is the large green *Manduca sexta* caterpillar, which is more commonly known as the "tobacco hornworm." Because of its green color, this caterpillar is often overlooked by predators when hidden in a plant's leaves, but the wasp hunts by smell, and the smell of the caterpillar's feces gives it away. This smell guides the wasp to its prey. It then lands on the back of the caterpillar, punctures its hide, and injects as many as several hundred eggs into the victim's body, using its tubelike stinger. A few days later, wasp larvae emerge from the eggs and begin to grow. As the larvae near maturity, they chew their way through the body of the dying caterpillar and spin cocoons on its back.

However, there is a danger that the caterpillar's potent defenses could kill the wasp's offspring, so the wasp has to disable them. Furthermore, the wasps must do this without killing the caterpillar, otherwise the wasp larvae will die too. Therefore, the wasps need to keep the caterpillar alive but defenseless. To achieve this, when laying her eggs, the wasp injects the caterpillar with venom, which includes a debilitating virus. This virus produces toxic proteins, which impair the caterpillars in a number of ways. The most important of these is to disable the caterpillar's immune system and hamper its physical growth, making it easier for the wasp larvae to take over. Gross but clever!

is it true that a cockroach farts every fifteen minutes?

There are more than 3,500 types of cockroaches of all shapes and sizes. The "hissing cockroach" from Madagascar is as big as a mouse, while the smallest cockroaches are the size of a tomato seed. Until recently, a beast called the "giant burrowing cockroach" held the record for the world's largest roach. This wingless cockroach, which bizarrely is a popular pet in Australia, grows up to 3.5 inches (9 cm) long—about the size of an adult finger—and weighs more than 1.5 ounces, which is about the weight of an AA battery. Despite its large size, it's able to flatten its body so it can fit into the smallest spaces. However, in 2004, scientists exploring caves in Borneo discovered a new species of cockroach. The monster roach measured 4 inches (10 cm) long, and this new discovery has taken over as the world's largest roach, in terms of size, although the giant burrowing cockroach is still the world's heaviest.

Cockroaches will eat just about anything, and they have teeth in their stomachs. They will happily eat the toe jam on your feet, the moisture in your nostrils, the sweat in your armpits, and even their own skins or other cockroaches; they really aren't fussy. The American cockroach, *Periplaneta americana*, is a large brown winged cockroach, about 1.5 inches (4 cm) long. It is commonly found in the southern United States and in tropical climates, and will often be found living in sewers. It will eat practically anything, including leather, bookbindings, soap, glue, and flakes of dead skin. It has also been known to munch on the eyelashes, eyebrows, fingernails, and even toenails of people while they sleep. Grossly, as cockroaches eat their food, they also poo on it and then trample over their next meal. All sorts of harmful germs are found in their feces and can cause illnesses in humans, such as diarrhea, typhoid, lung problems, and urinary tract infection.

Cockroaches also have other gross habits—they love to fart, and on average break wind every fifteen minutes. They also continue to release stinky methane gas for eighteen hours after they die. Insects called termites also fart a great deal and, along with cockroaches, are believed to be among the biggest contributors to global warming.

Cockroaches are also some of nature's hardiest creatures. Some species of cockroach can survive in blazing heat, while others can endure being frozen, then thawed, and still crawl away unharmed. Some cockroaches can go without food for three weeks. This astonishing creature can even survive for up to nine days after being decapitated, as it can breathe through a series of tubes attached to small openings on its body, which allow oxygen to pass to its cells. However, without a head it eventually dies from starvation!

what caused members of parliament to flee the houses of parliament in 1858?

In the 1850s, London was a pretty disgusting place. The streets, streams, and rivers were choked with sewage, and much of the waste from sinks and toilets ran down old sewers straight into the River Thames. Since most Londoners also got their drinking water from the Thames, it's hardly surprising that many people became sick and even died as a result.

During the long, hot summer of 1858, people of central London were plagued by a foul stench. London at this time was nicknamed the Great Stink, and its people were called the Great Unwashed. It was greatly overpopulated, and the sanitation was poor. The summer was so dry that much of the water in the Thames evaporated, and so river levels became very low. This meant that sewage dumped into the river was not carried out to sea, so it became an open sewer. For years, human sewage, slaughterhouse waste, rubbish, dead animals, chemicals from factories, and even dead bodies had been dumped in the Thames. Unsurprisingly, the stench became unbearable and the disgusting odor could be smelled from sixty miles away.

As the river's fish died out, the riverbanks became covered in a purple-pink mass of squirming tube worms. Sanitation workers deposited tons of lime into the river to help combat the foul stench, and lime-soaked curtains were hung in the windows of Parliament to mask the smell, but eventually the stink got so bad that members of Parliament fled from the Houses of Parliament. Heavy rain finally broke the hot weather and restored the river to normal levels.

which animal removes its own stomach to feed?

The common starfish (*Asterias rubens*) is not actually a fish but an echinoderm, which is closely related to other species such as sea urchins. The common starfish can grow up to 3 feet (1 m) across and has five arms, which contain tiny suckers, and a central mouth underneath the upper body. However, there are some species that have as many as forty arms. Starfish mostly live in the Atlantic Ocean but are also found in the Mediterranean.

Starfish feed on clams, mussels, oysters, and worms, and when they've found their food, such as a clam, they put their arms around it and grip it with their suckers, which gradually pulls the victim's shell apart. The starfish then pushes its whole stomach out of its mouth and into the shell of its prey. Once inside, the stomach releases digestive juices and begins to slowly digest the clam. When it has finished its meal, it sucks its stomach back into its body through its mouth.

Starfish are extraordinary creatures in many ways. When they are attacked, some may lose an arm or two; however, this isn't a problem because they can simply regenerate new arms. Most species must have the central part of the body intact to be able to regenerate, but a few can grow an entire new starfish from a single arm. Before this was known, fishermen would cut starfish into pieces thinking it would kill them, but the starfish grew into new ones, and so their population increased dramatically.

To reproduce, males and females release clouds of sperm and eggs respectively into the water, and fertilization takes place there, which is something to bear in mind when swimming in the sea.

why do male hippos shower each other in feces and urine?

Most of the world's hippopotamuses are found in Africa and can weigh as much as 3.5 tons (3.2 MT). They spend most of their time in water, where they defecate in copious amounts. The dung is important for the food chain, as it contains important nutrients for many microorganisms, which are in turn eaten by fish.

Around the Nile, male hippopotamuses mark off their territorial boundaries by performing a disgusting ritual. As the hippo walks, it defecates, and as it does so, it spins its tail to distribute the excrement over the largest possible area. Hippos also urinate backward, probably for the same reason. At night, they leave their rivers and lakes to graze on grasslands. To ensure they can find their way back to the water, they mark their trail by leaving piles of dung; the smell helps them to follow their trail.

Also, to indicate their status, dominant males will stop and stare at each other, and then turn backside to backside and shower each other in urine and feces, propelled with the help of their paddling tails, before walking away from each other. This behavior may seem gross, but it does at least provide a snack for nearby baboons.

which frog gives birth through its mouth?

Frogs tend to lay a large number of eggs to maximize the odds of survival because there are many hazards between an egg being fertilized and becoming a full-grown frog. However, some frogs have some very unusual ways of helping to ensure that their offspring get the best possible start in life.

Male Darwin's frogs (*Rhinoderma darwinii*) attract the female frogs by singing to them. After this romantic courtship, the female leaves the male to watch over the eggs. He gathers them up and swallows them, storing them in his vocal sac. Here, the eggs develop into tiny frogs before being spat out into the world.

Another amazing frog was the Australian gastric brooding frog (*Rheobatrachus silus*), which sadly was only discovered just before it became extinct. This fascinating creature would lay its eggs in the water, where they would be fertilized by the male and then swallowed by the female. Inside the female's stomach, the eggs would develop into tadpoles before emerging five weeks later as baby froglets from the mother's mouth.

But perhaps the most bizarre frog reproductive method is that of the brightly colored strawberry poison-dart frog (*Dendrobates pumilio*). This amazing tiny frog first lays five to ten eggs on a horizontal leaf, which both the male and female visit regularly, but the male has the job of keeping the eggs moist by urinating on them until they hatch. Then, when the eggs develop into tadpoles, the mother places them singly into small pools of water (because they are cannibalistic) and provides them with eggs to feed on that have failed to fertilize.

why do herring fart?

Some fish, such as the male cod, are known to produce a kind of grunting or buzzing noise to attract females, but these noises are not really farts, as they are not produced by the fish's anus. Instead, the male cod makes this noise by vibrating an air-filled sac called a "swim bladder," which lies beneath its backbone and is used to control the fish's buoyancy.

Another flatulent sea inhabitant is the Sand Tiger Shark, a ferocious-looking fish that has sharp teeth that protrude in all directions, even when the mouth is closed. The shark is denser than water and lacks a swim bladder, so it gulps air into its stomach at the surface of the water, then lets the air out as gas to help itself sink deeper into the water. This also allows the shark to float motionless in the water so it can seek prey. However, unlike human farts, these specialized emissions aren't produced by gas after eating.

One fish that does fart is the herring, which does so in a fascinating way and which seems to be unique to this species. It seems that herring make farting sounds to let one another know where they are, and these noises are so high-pitched that only other herring are able to hear them. No other fish has been known to make noises from its anus, or to produce such a high-pitched noise.

"It sounds just like a high-pitched raspberry," says research leader Ben Wilson of the University of British Columbia in Vancouver. Wilson and his team cannot be sure why herring make this sound but believe that it might explain the enigma of how herring shoals stay together at night. Originally, biologists thought that the high-pitched sounds were produced by the herring's swim bladder. However, they then noticed a fine stream of bubbles from the fish's anus always accompanied the farting noise. "In video pictures we can see the bubbles coming out of the anal duct at the same time," said Robert Batty, senior research scientist at the Scottish Association for Marine Science in Oban.

The team, who clearly had a sense of humor, called this noise Fast Repetitive Tick, or FRT for short. Wilson remarked that the number of FRTs produced does not change when the fish are fed. Also, if the fish were starved, they still produced the sounds. One other theory was that the FRTs were produced out of fear, but this too was rejected. The herring did not appear to be farting out of fear, because when they were exposed to shark scent, there was again no noticeable increase in bubbles or sound.

The researchers eventually became convinced that the FRTs were most likely produced for the purpose of communication, for three reasons. First, they noticed that more FRTs were produced when the herring were in bigger shoals. Second, herring are noisy only at night, so it has been postulated that rather than fumbling around in the dark, the fish can easily locate one another using their FRTs. Third, the FRTs have the additional benefit of being a safe way for the herring to communicate, as only other herring can hear them, so this method carries no risk of alerting predators to the herring's position.

who was mike the chicken?

We often use the phrase "running around like a headless chicken," and it's a well-known fact that when chickens are beheaded they may frantically run around for a time. However, a chicken named Mike lived perfectly well without a head for eighteen months, after a failed attempt to kill him in 1945. After his head was cut off with an ax, Mike continued to run around, and generally lived as full a life as a headless chicken might hope for. He preened and pecked with his neck, and was fed with ground-up grain and water through an eyedropper.

Mike is believed to have survived so long because much of his spinal cord and brainstem remained intact. He even became a national celebrity, touring the country and being featured in newspaper and magazine articles. However, one tragic night, Mike the Chicken choked to death.

In remembrance of Mike's life, the people in Fruita, Colorado, celebrate a yearly event every May called "Mike the Headless Chicken Day." This celebration includes events such as "Run Like a Headless Chicken Race," "Egg Tossing," and "Chicken Dance Contest." A website exists in tribute to Mike's memory at www.mikethehea dlesschicken.org.

can it really rain frogs and fish?

Surprisingly, it is not only frogs and fish that have been reported to have fallen from the skies, but also showers of snails, maggots, worms, snakes, shellfish, and even lizards. There have been thousands of reports from all over the world of showers of either small fish or tiny frogs raining down. However, strangely, nobody has ever seen frogs, fish, or other animals being carried up into the skies. The only rational explanation for these bizarre rains is that tornados or water spouts are responsible, as they tend to suck up whatever is in their path and, if the objects are light, take them into the storm clouds. A tornado can carry objects for several miles before eventually dropping them back to earth.

Other notable documented events of strange rainfalls include:

In A.D. 77, the Roman historian Pliny reported a shower of frogs that fell from the skies. In the fourth century, fish reportedly fell on a town in Greece for three whole days.

On April 24, 1871, during a heavy thunderstorm, a large number of sticky drops of jelly fell at a railway station at Bath. According to *Symons' Meteorological Magazine* (1871), the egg masses hatched quickly into annelid-like worms. A local vicar examined the jelly and watched them turn into some kind of larvae, which he described as "minute worms in filmy envelopes." In 1894, another bizarre shower fell on Bath, which was described in *Notes and Queries* (September 8, 1894) as consisting of thousands of shilling-sized jellyfish.

On December 15, 1876, live, dark brown snakes, from 12 to 18 inches (30.5 to 46 cm) long, covered the ground after a torrential rain on Memphis, Tennessee.

During a summer storm in 1894, a mass of small frogs rained down on Wigan in Lancashire.

On June 24, 1911, at Eton, Bucks, the ground was found covered with masses of jelly the size of peas after a heavy rainfall. Apparently, these lumps of jelly contained numerous eggs, from which larvae soon emerged. It was reported in *Nature* (July 6, 1911) that they yielded larvae of a species of midge.

During a storm in England in 1939, so many frogs fell from the skies that witnesses were afraid to walk around for fear of squashing them.

Perhaps the most disgusting of all these strange events was that which occurred in Bucharest, Romania, on July 25, 1872. On this day, the heat was stifling and the sky cloudless. At about nine o'clock, a small cloud appeared on the horizon, and a quarter of an hour afterward, to the horror of everybody, there was no rain, but small, fat, black worms or grubs covered most of the streets.

which animal licked leaking fluid from henry viii's rotting corpse?

King Henry VIII (1491–1547) ruled England in the sixteenth century. Henry suffered poor health in his later years, and in the late 1520s he began to suffer from leg ulcers, which caused him a great deal of discomfort and may have been the result of varicose veins or an infection of the bone. When the ulcers swelled he suffered with pain and dangerous attacks of fever. He is reported to have smelled awful, and needed constant medical attention.

Author Robert Hutchinson gives a description of Henry's awful disease:

> He is the personification of geriatric decay. One can almost smell the putrid stench of the rank pus oozing from his ulcers, staining the bandages on his swollen legs. Chapuys [the Spanish ambassador] labeled them "The worst legs in the world."

After Henry's death in London in 1547, his huge corpse was put into a large lead coffin, and resting on the coffin was a lifelike wax effigy of the king dressed in velvet. There was a procession through the streets of London, and the coffin was put down in Syon Chapel—which is where things started to become really unpleasant. One grisly story recalls:

> The king, being carried to Windsor to be buried, stood all night among the broken walls of Sion, and there the leaden coffin being cleft by the shaking of the carriage, the pavement of the church was wetted with Henry's blood. In the morning came plumbers to solder the coffin, under whose feet was seen a dog creeping and licking up the king's blood.

do jackals really feed their offspring with their own vomit?

The jackal, a doglike creature with a bushy tail, is found in many parts of Africa, the Middle East, and India. Jackals are members of the dog family, and can actually interbreed with both domestic dogs and wolves. Their predators include leopards, hyenas, and eagles. The ancient Egyptians believed a jackal-headed god, Anubis, guided the dead to those wise beings who would judge their souls.

Jackals are scavengers that find and eat dead animals in order to survive. They will happily eat decomposing or diseased flesh riddled with maggots, even if it's been rotting for days. When lions and tigers are done with their kills, jackals will happily move in for the revolting leftovers.

Unlike many other animals, jackals mate for life, producing many offspring. After hunting for food, the parents swallow prey they have caught, and when they return to the den, the pups lick their faces until the parents regurgitate the softened food. This rotting, regurgitated mush is then fed to the young!

what is a "sea cucumber," and why does it eat feces?

"Sea cucumbers" (*Holothuroidea*) are echinoderms, which means they have spiny skin. There are more than one thousand species of sea cucumber, many of which are, as their name suggests, shaped like soft-bodied cucumbers. All sea cucumbers live in the ocean, either on or in the ocean floor, and they can grow up to 3 feet (91 cm) long. They move sluggishly over the sea bed and often play host to a pencil-thin fish called a "pearfish," which lives in its body and moves in and out of the sea cucumber via its breathing hole, which is also its bumhole!

When confronted by a predator, such as a crab, some sea cucumbers, rather like Spiderman, discharge sticky threads to entangle their enemies, which buys them time to make a hasty escape. Others will violently contract their muscles, causing their internal organs to shoot out of their anus, to scare away the predator.

The sea cucumber's diet consists of feces and marine snow. The feces of many animals living in the ocean makes its way to the ocean floor. Marine snow is made up of mucus and small pieces of dead plants, animals, and bacteria. Both the feces and marine snow sink slowly through the ocean depths to the sea floor, where bottom-dwelling scavengers such as worms, clams, sea urchins, and sea cucumbers pick up and eat these nutritious foods.

Amazingly, despite the sea cucumber's diet, some cultures consider it to be a delicacy, and an aphrodisiac too—perhaps because the males shoot sperm into the ocean to reproduce. For centuries, the sea cucumber has been highly regarded for its therapeutic and culinary properties by the Chinese.

why do dogs eat poo?

We all love dogs, don't we? They're so cute and playful and affectionate. They wait at the door for you to come home from work, and pine when you leave. They'll catch a Frisbee, or scare off a burglar. They'll guide the blind, sniff out drugs and bombs, and rescue stranded explorers. Dogs are wonderful animals. If only they weren't so, well, disgusting. . . .

For a start, dogs will eat or drink almost anything: stagnant pond water, rubbish, vomit, cat poo, other dogs' poo, even dead, maggot-ridden roadkill—and then lick your face. And they seem to be particularly keen on cat poo. Why is this? One reason may be that cat poo is full of things that dogs like and need, such as protein. Protein is needed by animals and humans to keep healthy—fortunately, we humans find other ways to get protein, such as by eating meat and cheese.

Other unpleasant canine habits include licking their bums and dragging them across the carpet. The reason they do this may simply be that they are suffering from blocked anal glands, which their licking and chafing may be designed to unblock. A dog's anal sacs are found on either side of its anus and contain a stinky, fishy-smelling liquid, which dogs use to mark their territory. Some poor pooches suffer from trouble with their anal sacs, which can become blocked and irritated. These sacs, when full, are about the size of a marble. Each anal sac has a small tube that opens out to the bumhole. Every time a dog poos, some of this liquid is squirted out at the same time. It gives each dog's poo its own unique aroma, which tells other dogs to move their furry butts, because this territory is taken.

which centipede preys on bats in midflight?

Many people are grossed out by insects, especially giant ones, such as the Amazonian giant centipede (*Scolopendra gigantea*). This maroon-red centipede has forty-six banana-colored legs, and adults can reach the whopping length of 12 inches (30.5 cm). These creatures are aggressive, extremely fast runners, and excellent climbers.

The giant centipede usually feeds at night, when it creeps out to stalk its unsuspecting targets, which can include birds, lizards, frogs, and mice. It gropes through the darkness using its long antennae before expertly and quickly snagging its prey. It uses its jaws and modified front claws to inject its prey with a powerful and toxic venom. After a quick fight, the quarry dies and the centipede can then tuck into its kill. The giant centipede's venom is highly toxic to humans and can cause fever, pain, vomiting, and severe swelling; it has even been fatal in some cases.

However, the giant centipede's most extraordinary talent is the one it exhibits while lurking in the dark caves of Venezuela, where it hunts for bats. First, the centipede climbs over numerous writhing beetles and then up the wall of the cave. It then scurries across the ceiling into a position near the center. There, it grips to the roof of the cave with its rear legs, allowing the front part of its body to dangle. As the bats fly through the cave, unaware of the danger, the centipede grabs one from the air in midflight. The giant centipede's toxic venom works quickly, and although the bat fights valiantly to get away from its many legs, the struggle proves futile, and moments later it succumbs to the venom. While still hanging from the ceiling, the centipede devours its kill.

what is dangerous about the "silver carp"?

In the 1990s, a type of fish called the "silver carp" first entered the Mississippi River after having escaped from a local fish farm following a series of floods. Since that time, the carp have steadily moved upriver, and now they outnumber local fish by ten to one and are spreading to other rivers in the area.

One unique danger posed by these fish is the fact that they can leap out of the water; in fact, some have been known to jump 10 feet (3 m) in a single leap. It seems that one thing that makes them jump is the noise of boat engines—presumably because they think it is a predator chasing them—and this makes them especially dangerous, from a human point of view, as it means they literally jump with fright at every passing boat. If it were only a handful of fish jumping at any one time, that would be dangerous enough, as they can reach up to 100 pounds (45 kg) in weight—people have described being hit by a silver carp as like being hit by a bowling ball. In fact, however, they are known to leap in great shoals, with as many as two hundred leaping out of the water simultaneously, making them an even more serious menace.

There have been numerous reports of large jumping silver carp severely injuring boaters and water skiers. In June 2004, a woman water-skiing in Illinois was smacked in the head by a massive leaping carp and suffered a concussion. In another case, a fisherman had a big carp suddenly launch itself into his groin. Needless to say, he was in a lot of pain, but thankfully he did make a full recovery.

which spider carries its babies on its back?

If you take a walk through long grass in the summer, you may notice lots of small spiders scatter from your path. Many of these brown-colored spiders will be wolf spiders, which are unusual in that they do not make webs to catch their prey. Instead, they hunt, rather like wolves, which is the reason why they are so named. Wolf spiders have excellent vision, with eight big eyes that allow them to be very effective in spotting prey, which they then chase down and pounce on. The wolf spider then injects venom into its victim through its powerful fangs, quickly paralyzing or killing it. When the prey no longer struggles, the spider injects digestive fluids, which turn its prey's insides into liquid. The victim is then sucked dry of all its nutrients. Wolf spiders hunt small insects, and sometimes even other smaller wolf spiders. Worldwide, there are more than two thousand species of wolf spider, the largest almost 2 inches (5 cm) long, including their legs.

Rather like a man giving a bunch of flowers to the woman he loves, a male wolf spider may court the larger female by offering her a silk-wrapped dead fly. If successful, he will then transfer his sperm to her via his syringe-like feelers. In other species of wolf spider, when a male spots a female, he performs a courtship dance that involves waving his front legs like signal flags and vibrating his abdomen. The female is often not impressed by this showmanship and will react by attacking him. However, if the male persists, the female may allow him to climb on her and mate.

After mating, the female creates an egg sac and attaches it to her abdomen so that she can carry her eggs around with her. When the many spiderlings have hatched, the female wolf spider will carry them on her back for a week or more. In some species, the brood is so numerous that the spiderlings cover the mother's back several layers deep.

did little miss muffet really eat mashed-up spiders?

Little Miss Muffet
Sat on a tuffet,
Eating her curds and whey;
Along came a spider,
Who sat down beside her
And frightened Miss Muffet away.

The origin of this nursery rhyme is uncertain. However, "Little Miss Muffet" first appeared in print in Scotland in 1805, although it was probably around for a lot longer. Some historians believe that the nursery rhyme is based on a little girl called Patience, whose father was an eminent English doctor called Thomas Muffet (1553–1604).

Thomas Muffet was passionate about bugs, especially spiders, and how they related to medicine. He was a strong believer in the power of spiders to cure nearly any ailment. Most of his remedies consisted of eating spiders, but some were more imaginative. For example, one of Dr. Muffet's recommended treatments for cuts or abrasions was to wrap the wound in fresh spiderwebs, which apparently worked quite well.

He is thought to have instructed his daughter to eat mashed spiders, believing that they would help her recover from colds—in fact, this was a fairly common remedy for colds two hundred years ago—and so Dr. Muffet and his daughter Patience may be the origin of the nursery rhyme.

why should you be wary of a "brown recluse"?

Worldwide, there are more than 34,000 species of spiders, and most of them are venomous. One of the most dangerous spiders in the United States is the "brown recluse" spider (*Loxosceles reclusa*), which is also known as the "fiddleback spider," "brown fiddler," or "violin spider," on account of a distinctive mark on its back that is in the shape of a violin. One bite from a brown recluse can cause a range of extremely unpleasant symptoms, and can even be fatal.

The brown recluse spider is usually between 0.2 and 0.7 inches (5 and 18 mm) long, and it ranges in color from tan to dark brown. Besides the violin marking, it can also be identified by its unusual number of eyes. Most spiders have eight eyes, but the brown recluse has only six, which are arranged in three pairs. The brown recluse is native to the United States and is generally found between the southern Midwest states and the Gulf of Mexico. It usually builds its webs in sheds, garages, and cellars, and other places that are dry and undisturbed.

As its name suggests, the brown recluse is rather timid and hides for most of the day but it is active at night, and so is more likely to find its way into clothing, bedding, and shoes at this time. It would rather run from danger than fight, but if cornered, it will bite, and its venom, sphingomyelinase D, is responsible for an array of gruesome symptoms. People rarely see the spider, and so they are often unaware that they have been bitten until the symptoms begin. The severity of symptoms depends upon the amount of venom injected, and the individual's sensitivity to it. After being bitten, some people will experience a stinging sensation in the bite area. Within several hours the individual may suffer with increasing pain and itching. Also, the area may become red, swollen, and tender, and an unpleasant white blister may form in the center of the wound.

Most brown recluse bites are fairly minor, but some wounds may become "necrotic," which is very nasty indeed. Necrosis, the root of the word "necrotic," means the premature death of living cells or tissues as a result of external factors, such as injury, toxins, or infection. If the wound becomes necrotic, within two to five weeks the spider's venom will start to destroy the tissues at the site of the bite, resulting in a deep, poorly healing ulcer. The spot will grow deeper and larger, and may expand up to several inches over a period of weeks. The victim's flesh will become hard and blackened. Other symptoms can include fever, chills, headache, nausea, vomiting, weakness, and restlessness, and full recovery can take several months, leaving a nasty scar. Deaths from brown recluse bites are rare, but there have been some among children, the elderly, and people with poor health.

which insect emerges from the ground in the billions every seventeen years?

A cicada nymph is a strange-looking insect with bulging eyes and big hooks on the end of its front legs. It can grow up to 2 inches (5 cm) in length, which means adult cicadas are among the biggest of all insects. They have two pairs of wings, broad heads, and thick, heavy bodies.

The nymphs live underground in total darkness, where they mature over a number of years. During their time underground, they have a ready supply of food, as they suck sap from the roots of trees. The

Magicicada septendecim cicadas live for seventeen years underground and, after this time, emerge from the ground in their billions in places such as Illinois, Wisconsin, Michigan, and Indiana. Exactly how the cicadas keep track of time while underground has always puzzled researchers. One group of scientists in California has suggested that the nymphs use an external cue, which they are able to count. They suggest that when the trees flower, this coincides with a peak in the chemical concentration of the sap that the insects feed on. So, perhaps, the cicadas may "tick off" the years by counting these peaks.

After their seventeen years underground, the cicadas finally emerge. When they reach the surface, they then find the nearest tree to climb and head for the treetops. Once the cicada has climbed a tree, the skin on its back splits and it slowly moves up and out of the slit, leaving its exoskeleton behind. Its crumpled wings now pump full of blood and harden, allowing the cicada to fly. Then, at last, the mating season can begin.

When they are mating, the male cicadas make a lot of noise, using a drumlike organ in their body. In fact, they are so loud, the combined noise made by the male cicadas is said to equal 100 decibels, which is louder than a jet aircraft flying overhead, and can be heard a quarter of a mile (402 m) away.

Once the cicadas have mated, the male soon dies. The female lays hundreds of eggs, which she does by making incisions in the bark of a twig and then laying the eggs inside the resulting slit. Once this is done, she too drops dead. The sudden deaths of all these cicadas leaves 100 tons of dead and dying carcasses to be cleared up from the area.

After leaving the egg, the cicada nymph emerges from the twig and falls from the tree to the ground. It then begins crawling around, looking for a crack in the ground or some other suitable place to dig, and hollows out a small chamber around itself. Once the nymph has found a good spot, it will puncture the root of a nearby tree with its sharp beak and begin sucking the tree's sap, and this is where it will stay for the next seventeen years.

what is "monkey-face" lamb disease?

Who would think a pretty, delicate white-flowered plant could cause such an unsightly deformity as a lamb born with a single eye in the middle of its forehead? Malformations such as these can result if pregnant ewes eat the poisonous plant *Veratrum californicum*, which is commonly known as "false hellebore" and can be found on the high mountain ranges of the northern Rockies and in southwestern North America.

The most common malformation caused by false

hellebore is called "monkey-face" lamb disease, in which the lambs are born with deformed heads. Usually, the lamb's nose will be shortened, or gone altogether, and its jaw will protrude. The lamb may have an enlarged head, and its face may be caved in. In extreme cases, the lamb might even have both eyeballs in one eye socket, in the middle of its forehead, creating a cyclopic lamb. Unfortunately, although most livestock generally avoid false hellebore, sheep do not.

If you ever find yourself in Paris and have an urge to see a cyclopic head, why not head for the Fragonard Veterinary Museum? There you will find a wide range of exotic and unusual animal parts on display. At this fascinating museum, not only can you view a cyclopic veal calf's head floating in a jar, but you might also like to observe a cow's ovarian cyst, a horse's tongue, or perhaps a skinned cat stretched out like a little rug.

which insect is the tarantula's worst enemy?

The tarantula may seem like a fearsome predator, but it's no match for its worst enemy, a wasp called a "tarantula hawk" (*Pepsis heros*). The tarantula hawk wasp is one of the largest types in the world, with a wingspan that reaches 4 inches (10 cm) and a body length of up to 3 inches (7.5 cm). It is found worldwide and is typically metallic blue or black with brightly colored wings.

When the female wasp picks up the scent of a tarantula, she will hunt it down and then attack it. Strangely, the tarantula rarely attacks the wasp and becomes quite docile. Some experts think the wasp produces a pheromone—a chemical that causes the tarantula to become stupefied. The wasp will then sting the spider, quickly paralyzing the arachnid with its neurotoxic venom. The female wasp then lays an egg on the spider's motionless body and pushes it into a burrow. When the hungry baby wasp hatches, it has a ready supply of fresh meat on which to feed. At first, it plunges its mouthparts into the spider's body and feeds on the spider's juices, but as the larva grows quickly, it soon begins eating a diet of solid foods—the spider's organs. All of this happens while the spider is still alive.

what is the schmidt sting pain index?

If you've ever suffered the pain of a bee or wasp sting, it can be hard to imagine why someone would encourage insects to sting them. However, one brave entomologist decided to allow a variety of insects to inflict stings on his body, all in the name of science.

In the 1980s, biologist Justin O. Schmidt of the Southwestern Biological Institute in Tucson, Arizona, devised a scale of insect stings according to how much they hurt. The "Schmidt Sting Pain Index" is not a scientific scale, but it does give us some idea of which insects have the most painful stings. The index starts from a "0," which is the sensation of being stung by an insect that cannot penetrate human skin, progressing up to a "4," for the most painful stings.

According to Schmidt, the most painful stings come from tarantula hawk wasps, bullet ants, velvet ants, and pepsis wasps. He described the tarantula hawk's sting as being "blinding, fierce, shockingly electric. A running hair dryer has been dropped into your bubble bath. If you get stung by one, you might as well lie down and scream." Apparently, the only insect with a worse sting is the bullet ant, a terrifying insect from the Amazon rainforest whose sting Schmidt describes as "pure, intense, brilliant pain. Like fire-walking over flaming charcoal with a three-inch rusty nail in your heel (Justin O. Schmidt, "Hymenoptra Venoms: Striving Toward the Ultimate Defense Against Vertebrates," 1990)."

which wasp lays her eggs inside a cockroach?

The "emerald cockroach wasp" (*Ampulex compressa*) is about 0.7 inches (2 cm) in length and is a myriad of metallic blues and greens. It is native to the Pacific Islands, Southeast Asia, and parts of Africa. When it finds an unsuspecting cockroach, it will sting it twice. The first sting is designed to paralyze the cockroach but not kill it. This paralysis is only temporary but gives the wasp enough time to carefully administer its second sting, because this time the sting needs to be much more precise. The second sting delivers venom into the cockroach's brain, affecting its escape reflex, so that when the cockroach comes out of its physical paralysis, it does not feel the need to escape. Instead, it simply grooms itself, giving the wasp time to look for a lair.

When the wasp returns, having found a suitable spot, it bites off one of the cockroach's antennae and laps at the resulting blood. Then it drags the cockroach around by its other antenna, like a dog on a lead, and takes it back to the nest. Then it lays an egg on the stupefied cockroach before burying it, to stop other predators from eating it. The cockroach is still alive but still has no instinct to escape, thanks to the wasp's earlier venomous injection. When the eggs hatch, the wormlike wasp larvae will begin to eat the living flesh of the cockroach. Eventually, after days of voracious eating, the helpless cockroach dies.

what is the "fiery serpent"?

The "fiery serpent" is the colloquial name for the "guinea worm" (*Dracunculus medinensis*), also known as "dracunculiasis," which is a very unpleasant parasite that infests humans. The guinea worm is most commonly found among poor communities in remote parts of Africa that do not have access to safe drinking water. It is renowned for causing persistent, excruciating pain and suffering. The worm begins life as a small larva, which gets swallowed by tiny water fleas in contaminated water.

For humans, the trouble begins when we drink water contaminated with these water fleas. Although gastric acid in the stomach kills the water fleas, it doesn't kill the guinea worm larvae, as they are resistant to this acid. Once released, the larvae then travel along our small intestine and begin their migration. They bore their way out of the intestine, through the intestinal wall, and during the next ten to fourteen months, they grow to adulthood and steadily make their way to our legs. As they make this journey, the worms also mate. After they have mated, the male dies and is absorbed by the female. All this time they are continuing to grow, and they can eventually reach as much as 3 feet (91 cm) in length—eventually, they end up looking like a very long strand of spaghetti.

At this point, the adult female steadily moves toward the surface of the skin; usually on the legs, but she may emerge from other areas. As the worm reaches the skin's surface, it creates a blister, causing intense, burning pain, before finally rupturing within about twenty-four to seventy-two hours. This may also be accompanied by fever, nausea, and vomiting.

To relieve the burning pain, many victims will usually try to immerse their leg in cold water, often in a river. However, the temperature change will often cause the skin over the blister to crack, exposing the adult worm and allowing it to release millions of young larvae into the water. At some point, many of these larvae will be eaten by passing water fleas, and the whole cycle will begin again. It takes about a year from the point a person drinks infected water for the worm to emerge from the skin.

However, the really unpleasant part of this story concerns the removal of the worm, as there is no drug to treat it. Our natural temptation is to grasp hold of the protruding end of the writhing worm and pull it out. However, this temptation must be resisted because it could be fatal. If a sufferer tries to rip the worm out, and in the process breaks its body, the part of the worm left under the skin will die, releasing large amounts of proteins that cause a sudden, lethal shock.

The only way to remove the worm safely is to pry it out very slowly, a few centimeters each day. This exceptionally slow, painful process is often done with a small stick about the size and shape of a pencil, and may take a few days to pull out, although sometimes it can even take weeks. Even once the worm has been successfully removed, people are often left with scarring, or are even permanently crippled. Thankfully, this dreadful disease is close to being eradicated, thanks to international efforts to ensure safe drinking water in the poorest parts of the world. According to the World Health Organization, in the 1980s, there were more than 3.5 million people infected with guinea worms throughout Africa and Asia. However, in 2008, there were only 4,619 cases, all of which were in Africa.

which worm can eat its way through the human eye?

There is a parasite called the "toxocara worm," which is commonly found in dogs and cats. It is the most common parasite in young puppies but causes them no problems. However, the worm can be a problem if transmitted to humans, which can easily occur. The toxocara worm produces a large number of eggs in the stomachs of cats and dogs, and these eggs are excreted with the animals' poo. All it then takes is for a person to tread on some dog poo in the street, and then transfer the eggs to the carpet back at home, or the garden, and it's easy to see how a person might become infected.

However, once inside the human, the toxocara worm does not tend to like this environment, so it will try to find its way out. This can result in affected parts of the body becoming inflamed, and, although this is rare, the worm may even emerge through the eye, the brain, or the heart. If the worm tries to emerge through the eye, it can ultimately lead to blindness. Treating ocular toxocariasis can be difficult; in some cases, a laser has to be used to kill the worm before it reaches the retina.

Nonetheless, most people who develop toxocariasis recover without needing treatment. Many of them suffer no symptoms at all, and never even realize that they had the infection. The good news is that even if symptoms do appear, antiparasitic medicines can be taken to destroy the worms. About 20 percent of adult dogs carry toxocara, along with around 80 percent of puppies, which is why it is so important for dog owners to deworm their pets.

how did the "death watch beetle" get its name?

The "death watch beetle" (*Xestobium rufovillosum*) is a small wood-boring beetle that is about 0.3 inches (7.5 mm) long. The male death watch beetle attracts a female by knocking his head against the wooden beams of the old building in which the female lives. She, in turn, answers by banging with her head.

For centuries people were terrified by the sound of the death watch beetle. In olden days, when people heard the ominous "tap, tap, tap" sounds, they thought it meant that someone in the house was going to die, which is how the beetle got its name. Many writers have made use of the death watch beetle's ominous, atmospheric quality. For example, in Edgar Allan Poe's sinister story "The Tell-Tale Heart," the unnamed narrator hears the death watch beetle tapping in the walls while he watches his victim in his bedchamber.

what are the world's scariest creatures?

As is probably becoming clear, the natural world is full of disgusting, vile, nauseating creatures, but which is the most sickening? One contender must be the goliath tarantula, which lives in the jungles of South America.

Alternatively, some people are more repulsed by crustaceans. The world's largest crab is the "Japanese spider crab" (*Macrocheira kaempferi*). Although its body measures only about 12 inches (30.5 cm), its outstretched leg span reaches more than 11.5 feet (3.5 m)! It can be found in the northern Pacific Ocean, where it lives in deep water and thankfully never surfaces. It is adapted to withstand great water pressure at depths of 3,300 feet (1,005 m) or more.

Another intimidating sea dweller is the "colossal squid" (*Mesonychoteuthis hamiltoni*), which can grow to 39 feet (12 m) in length and could easily play a starring role in a horror film. It has huge, rotating hooks that grow from its suckers, an extremely large, sharp beak, and enormous bulging eyes, which are the largest eyes of any animal—the size of a dinner plate. In 2003, fishermen in the Ross Sea of Antarctica became the first people ever to see a colossal squid alive. It was feeding on fish caught on their fishing lines. It was almost 20 feet (6 m) long, about half its full, adult size.

Africa's "goliath beetle" (*Goliathus giganteus*) is one of the biggest and heaviest insects in the world and can be found in tropical rain forests, such as the Amazon. They are as big as a human fist and measure about 6 inches (15 cm) long and 2 inches

(6 cm) across, and their primary food source is feces. They are good fliers and make the sound of a low, helicopter-like whirr while flying. Sometimes African children like to tie their bodies onto string and watch them fly around in circles. Male goliath beetles have a horn-shaped structure on their heads, which they use to fight with other goliath beetles.

chapter three

vile bodies

what did queen victoria see floating in the river cam?

Today, the River Cam in Cambridgeshire is one of the most picturesque of English rivers. However, back in the 1800s, it wasn't an idyllic place to sit beside and eat a picnic, as it was rather smelly. All of Cambridge's sewage flowed into it, making it simply a large open sewer, and a pretty disgusting place to walk beside.

In 1843, Queen Victoria was walking beside the river with her companion, a university lecturer named Dr. Whewell, and as they crossed the bridge, the queen looked down into the water and asked, "What are all those pieces of paper floating down the river?" They were actually bits of poo-covered toilet paper floating on the water, but the quick-thinking lecturer replied, "Those, ma'am, are notices warning that bathing is forbidden."

how did napoléon's hemorrhoids affect the outcome at the battle of waterloo?

Napoléon Bonaparte was born in Corsica in 1769. He became a general in the French army in 1796, and, after staging a successful coup in 1799, he crowned himself Emperor of France in 1804. Napoléon is considered one of the greatest military commanders of all time, and until 1815 he never lost a battle at which he was present on the field. However, in 1815, Napoléon was defeated at Waterloo, in what is now Belgium, by the British and Prussian troops under the command of the Duke of Wellington and General Marshal Blucher. But how could hemorrhoids have played a part in this defeat?

Napoléon suffered from both constipation and very painful hemorrhoids, which are itchy, swollen veins in the region of the anus. Historians say that before every battle, Napoléon would survey the battlefield on horseback. However, before the Battle of Waterloo, he was suffering from a particularly painful and severe case of hemorrhoids, which prevented him from being able to mount his horse, as it caused him severe physical discomfort. Maybe if he had been able to inspect the battlefield, he may have won the battle.

There is also a theory that Napoléon suffered from scabies. Many famous paintings of Napoléon show him with his hand in his shirt—this may have been a fashionable pose at the time, or perhaps he simply had an itch that needed scratching? Scabies is a parasitic infestation of tiny mites that burrow under the skin, leaving behind a trail of white and black dots, which are their eggs and poo, respectively, and causing intense itching. Napoléon was often to be found soaking in hot baths, sometimes several times a day. A person with scabies might take regular hot baths to soothe inflamed, itchy skin. Napoléon's troops during the Russian campaign were thought to have been widely infected with scabies, and Napoléon is thought to have transmitted them to his wife, Josephine. Napoléon's doctor was the Frenchman Corvisart, who is considered to have been one of the most brilliant physicians of the early nineteenth century. Corvisart was apparently able to cure Napoléon of scabies by prescribing a topical mixture containing alcohol, olive oil, and powdered cevadilla (derived from a tropical lily that grows in Mexico and Central America).

why would a man volunteer to be castrated?

Many readers may wince to learn that through the years, there have been many examples of men volunteering for castration. Some have done so out of religious piety. George Rapp, for example, was one of the most famous castrates of the nineteenth century. Rapp founded the Harmony Society of Utopian Christians, in Pennsylvania. Rapp believed that sex was immoral, and strongly urged his community to live in celibacy. To demonstrate his devotion, Rapp took the extraordinary step of castrating himself, and he was also accused of castrating his son and several other men. Castration was not uncommon in the early 1800s, as it was often prescribed in the treatment of mumps and other diseases.

There has also been a long tradition of castration among male singers, known as "castrati," who were enormously popular in Italy between the 1500s and the early 1900s. Castrati were regularly employed as chapel singers in both Italy and Germany. The Catholic Church had for centuries banned women from singing during services, and with the high popularity for religious music, the castrati were in demand. The Italian and German castrati were men who had high voices like those of a boy soprano, but with the lung power of grown men.

Unfortunately, the only way to achieve this kind of singing voice was for a singer to have his testicles removed before he reached the age of puberty. The reason for this is that during puberty, a boy's testicles begin to produce increased amounts of the hormone testosterone, which causes the length of the vocal cords to increase by about 50 percent, which causes the voice to deepen.

With the birth of opera in about 1600, the demand for castrati increased enormously. Castrati were extremely popular among the operagoing public, being renowned for their vocal range, power, and lung capacity. A potential castrato would undergo a voice evaluation, and if it was decided that the lucky lad's voice had potential, his parents would arrange the castration, in a hope it would provide him with a good future. The doctor carrying out the castration would usually drug the child with opium or some other narcotic, before placing him into a very hot bath until he was nearly unconscious, before finally chopping off his testicles. Unfortunately, some boys died while undergoing this operation.

Those who did survive could never have children and were destined to have an unusually small penis. Despite this, some of the celebrated castrati were also well-known ladies' men. Although they were sterile, they were not impotent, so they were therefore able to get erections. In the absence of testicles, the adrenal glands, found above each kidney, produce male hormones in sufficient amounts to help produce an erection.

The last known castrato, Alessandro Moreschi (1858–1922), was a singer in the choir of the Vatican's Sistine Chapel and was known as "the angel of Rome." He was castrated in 1865, just five years before the practice was officially outlawed. He made a number of phonograph recordings of his voice in the early 1900s; unfortunately, the recordings were of such poor quality that they sound rather strange to the modern ear and do not give a true reflection of his amazing vocal quality.

what underwear blocks the smell of farts?

Let's face it, everyone farts, and studies have shown that when men and women eat the same food, women produce more concentrated gas than men—in other words, stinkier farts! Farts are caused by the food we've eaten and the air we've ingested. However, there are some medical conditions known to cause flatulence, too.

For example, Arlene Weimer of Pueblo, Colorado, suffers from a condition called Crohn's disease, which causes symptoms including inflammation of the digestive tract, diarrhea, and flatulence. Luckily, Arlene's husband, Buck, is the inventive type and decided to take action. In an article featured in the *Denver Post* in 2001, Buck said, "I'm laying in bed with her, sort of suffering silently, and out of the silence came determination. Something had to be done." He set to work to try to come up with a solution to the problem of his wife's flatulence.

Buck had the idea of using a filter but realized it couldn't be too big and bulky, and, of course, it would have to absorb any bad smells. He experimented with the filters used in the gas masks worn by coal miners—he put one into his pants, and although it didn't work perfectly, he knew he was on the right track. He carried on working on prototypes until, six years later, he came up with his brand-new invention: Under-Ease, which are airtight underwear with a replaceable charcoal filter that removes bad-smelling gases before they can escape. In this way, the pants are supposed to completely hide the smell of farts, by preventing the noxious gas from escaping. Unfortunately, however, they don't prevent the "parp" noise that accompanies the trumping. Naturally anxious that no one should steal his revolutionary new idea, Buck received a patent in 1998.

Buck's Under-Ease pants are made from a soft, nylon-type fabric, with elastic around the waist and legs. The removable filter, which looks a bit like a shoulder pad, is made of charcoal sandwiched between two layers of Australian sheep's wool. Buck explains how the filter mechanism works:

> The multi-layered filter pad traps the 1–2% of human gas creating the foul smell (mostly

> hydrogen sulphide), but allows the remaining non-smelling gas to pass through. It also allows the natural build-up of body heat to pass through. They can be worn anytime, anywhere—to work, social events, professional meetings, even on an airplane. And thankfully, the pants are machine washable.

The company's motto is "Wear them for the ones you love."

what is a "love bug"?

If you're suffering with intense, persistent itching in the genital area, there's a good chance you could be suffering from "love bugs," which are also known as "pubic lice," or "crab lice." The French have a much nicer term for these little pests, giving them the rather poetic name of "*papillons d'amour*," which literally means "butterflies of love."

Pubic lice are usually transmitted during close physical contact with an infected individual. The lice may be seen moving around in the pubic hair and are particularly active at night, which causes intense itching. They are well adapted to live in pubic, underarm, beard, and eyelash hair, which can become quite unsightly when peppered with lice eggs. However, the pubic louse strikes gold when it finds an extremely hairy man, as a hirsute host will provide it with a wide expanse to frolic around on and lay eggs.

The crab louse is so named because of its shape and powerful claws, which resemble a crab's pincers. It uses these claws to grasp our hair, and it can live for up to thirty days in the pubic hair. Female crab lice stick their eggs to hair shafts using a cement-like substance. The good news is that there is an effective treatment that can be applied to the affected area to cure this itchy affliction.

what was the great house of easement?

In the Tudor era, there were no flushing toilets. Instead, Henry VIII rested the royal posterior on a "close stool," which consisted of a bucket and water tank with a padded seat lavishly covered in black velvet. His groom-of-the-stool was even required to wipe the royal bottom.

However, Henry's courtiers used the malodorous communal facilities of the "Great House of Easement" at Hampton Court, a two-story facility that consisted of rows of oak seats with holes cut in the wood at two-foot intervals. Chutes made of bricks or stone would carry the waste down to a basin, which then flowed directly into the River

Thames. However, some of the more solid waste would collect in the brick chambers, which had to be regularly cleaned. This stinky task was given to "gong scourers," who were also known as "gongfermors" or "gong farmers." The gong scourers were often young boys, as it was their misfortune to be small enough to crawl along the drains.

is it true that ivan the terrible would often bang his head on the floor until it was bruised and bloody?

Ivan the Terrible (1530–84) became the ruler of Russia at the age of three, following his father's death. After his mother was murdered by poisoning, the rich noblemen, known as the "boyars," took over the leadership of the country. Murders and beatings were regular events at the palace, and the boyars killed four of Ivan's uncles and other important personages at the court—these killings were often witnessed by Ivan. While Ivan was young, the nobles treated him very cruelly; he was neglected, dressed in rags, and beaten and tortured. He took his frustrations out on animals and enjoyed pulling out the feathers of live birds, piercing their eyes, and cutting open their bodies. As he grew older, he took his frustrations out on people and became cruel and vindictive. When he was fifteen, Ivan was offended by a nobleman who was apparently rude to him, so he ordered the man's tongue to be cut out. Ivan suffered with paranoia and had a vicious temper, and it was said that when he became angry, he would foam at the mouth like a horse.

Ivan appointed military officers known as "oprichniki." They wore black uniforms and carried a dog's head on their horse's bridle and a brush fastened to their whip. These items symbolized the hunting down of Ivan's enemies and then sweeping them away.

In 1547, seventy-five citizens of Pskov complained against the governor of Pskov, which enraged Ivan so much that he made them undress and lie on the floor with their hands tied behind their backs. As they lay prostrate, boiling wine was poured onto their heads, and their hair and beards were set alight.

In 1570, Ivan and his oprichniki attacked the city of Novgorod, as he was angry that the city did not submit abjectly to his authority. For five weeks they tortured, raped, and murdered up to 60,000 people. It is said that so much blood entered the city's river that it didn't freeze over in winter for years afterward.

In 1575, Archbishop Leonid of Novgorod was accused of a whole array of crimes, including treason, buggery, bestiality, and keeping witches. Leonid's large fortune was confiscated by Ivan, eleven of his servants were hanged, and his "women witches" were dismembered and burned. Legend has it that Ivan ordered Leonid to be sewn into a bear's skin and thrown to a pack of hungry dogs. In the same year there was a famine, and Ivan invited some starving peasants to dinner. Instead of feeding them, he murdered them and threw them into the river.

Despite his shocking brutality, Ivan nonetheless considered himself to be a deeply religious man, and would pray for forgiveness for his sins while banging his head on the floor till it was bruised and bloody.

who was the "fart maniac"?

In the 1800s, a French baker named Joseph Pujol (1857–1945) took to the stage, often wearing clothing that exposed his bottom, and made a lot of money entertaining large audiences with various tunes, all produced by farting.

His stage name was Le Pétomane, which literally means "the fart maniac," and his remarkable talents included imitating various sounds with his farts and smoking cigarettes with his anus. Using a rubber tube, he was able to blow out a candle 12 inches (30.5 cm) away just by bending over and farting. Also, he could even insert a small flute and play a couple of tunes.

Pujol first realized his "gift" after swimming in the sea at Marseille; he dived underwater and held his breath, and suddenly he felt something cold flow into his lower bowels. As he ran out of the sea, water poured out of his anus. Later, he discovered that by controlling his abdominal muscles, he could suck lots of water into his rectum and squirt it out of his arse. Naturally, from water he moved to air.

In 1892, he entertained an audience at Paris's Moulin Rouge. He would begin with a series of quiet farts, "This one is a little girl, this the mother-in-law, this the bride on her wedding night [very quiet], and the morning after [very loud]. . . . This is the dressmaker tearing two yards of calico [he farted for about ten seconds]." He completed the performance by mimicking the sound of cannon fire, producing an impressively loud, booming noise.

Intrigued by his talents, medical experts carried out an investigation into how Pujol was able to use his body in such a way. In 1892, he was studied in at least two major medical journals. In a detailed investigation, Dr. Marcel Baudouin observed,

"In a state of repose, the anus shows no sign of abnormality,but is perhaps a little more dilated than usual. The sphincter is strong and elastic."

what were "waterloo teeth"?

At the Battle of Waterloo in 1815, such was the scale of loss of life that there were rich pickings for the British troops when the battle was over. The dead were stripped of anything of value—including their teeth. Such a vast number of teeth were extracted from the dead at Waterloo that for many years afterward, dentures were known as "Waterloo teeth." A tooth transplant involved taking a tooth from one person and transplanting it into the head of another person, and many people unwittingly wore teeth extracted from the soldiers.

Tooth transplants plummeted in popularity when it was discovered that infections such as syphilis could be transmitted in this way.

have people ever been allowed to expose their private parts in public?

These days, it would be a definite fashion faux pas to go out wearing clothing that exposed one's genitals. However, that wasn't always the case. Back in the 1300s, men wore a kind of tights called "hose." Hose consisted of two separate stockings, tailored from cloth, which would be laced onto the bottom of the wearer's tunic. As the hose covered only the legs, and men didn't wear underwear, this left a gap between the legs, but this was fine as long as tunics were of a "decent" length. However, over time there developed a fashion for tunics to get shorter and shorter. Eventually, the tunics became so short that they didn't adequately cover a man's groin, leaving his private parts often in view.

In 1482, King Edward IV passed a law that men had to cover their genitals, and, as a result, codpieces quickly became very fashionable. Originally, the codpiece was a triangular flap in the hose, which served the function of a modern trouser fly: it could be unbuttoned or untied to allow the wearer to urinate. However, the codpiece became larger over time to help accentuate a man's private parts, with some codpieces protruding as much as 5 inches (13 cm). Author Michel de Montaigne (1533–92) was not a fan of the codpiece and sniffed, "What is the purpose of that monstrosity that we to this day have fixed to our trousers, and often which is worse, it is beyond its natural size through falseness and imposture." Men were often accused of stuffing them to exaggerate the size of their genitals.

By the 1500s, codpieces were heavily padded so that they stuck out from the rest of the clothes and were decorated with ribbons, puffs, and jeweled pins. However, by the 1590s, codpieces had become passé.

where can you find a place called "bollock"?

Throughout England, there are numerous places with rude and saucy connotations, including "Grope Lane," "Juggs Close," "Crapstone," "Crotch Crescent," "Slutshole Lane," and "Titty Ho." "Scratchy Bottom" can be found in Dorset, while there is a "Lickey End" in Worcestershire. However, smutty place names are not confined to the UK. The following is a list of some of the more outlandish place names from around the world. . . .

Bastard, Norway
Bollock, Philippines
Bumbang, Australia
Chinaman's Knob, Australia
Condom, France
Crap, Albania
Dikshit, India
Dildo, Canada

Fucking, Austria
Knob Lick, Missouri, United States
Labia, Belgium
Minge, Lithuania
Semen, Bulgaria
Shag Island, Australia
Shit, Iran
Slut, Sweden
Turda, Romania
Vagina, Russia
Wank, Germany

did hitler really suffer from flatulence?

According to biographer John Toland, Adolf Hitler (well known for being a hypochondriac) suffered from uncontrollable farting. Hitler told a doctor in the early 1930s, “The cramps are so severe that sometimes I could scream out loud!” However, a Berlin doctor named Theodor Morell prescribed Dr. Koester’s Anti-Gas Pills, which Hitler took for many years. However, although they apparently succeeded in reducing the flatulence, it seems the pills would have done him more harm than good, as they contained strychnine and belladonna, which are two deadly poisons. According to Alan Bullock in his biography of Hitler, the poisons accounted for the führer's ash-gray skin coloring and shaking limbs toward the end of his life. Severe periodontal gum diseases also caused him to suffer with halitosis (foul-smelling breath).

which disease caused its victims' organs to liquefy?

The Black Death or bubonic plague (1348–51) was the most virulent plague that history has ever recorded. Its name came from the black swellings that were found on many of its victims, and other symptoms included pain, fever, and vomiting blood. The plague caused internal bleeding, and in the final days, the victims' vital organs began turning to liquid and rotting away. Mercifully, death would come soon after. Once a person contracted the plague, recovery was rare, although a few people did survive the disease.

The Black Death was responsible for killing up to half the people in Europe, and struck England in 1348. It was at its worst in cities, where there was enormous overcrowding and poor hygiene and sanitation. When the plague reached London, up to 30,000 people died, which was nearly half of the city's population. Over the period of the Black

Death, the population of England halved from around 6 million to 3 million. So many perished that the graveyards were soon full and corpses had to be buried in fields.

Plague epidemics occurred regularly across Europe following the Black Death. The last outbreak of plague in England was the Great Plague of 1665. As in 1348, the plague spread quickly and killed many people. The plague's victims developed painful pus-filled swellings, or "buboes," in their groins and armpits. The buboes were red at first but later turned a dark purple or black. When spots caused by blood seeping from damaged blood vessels beneath the skin were found on the body, this was the death certificate for almost all sufferers. The pain was so unbearable that some victims threw themselves out of windows. Sufferers also experienced an unquenchable thirst that led to some of them running naked through the streets screaming, and plunging themselves into water tanks.

To help ensure victims did not pass on the infection, they were forced to stay in their houses, and the doors were nailed shut and marked with a large red cross. "Searchers" were paid to remove corpses of victims from homes, and walked around streets calling, "Bring out your dead!" As more people became sick, the rich left the city and most of the doctors and clergy went with them. The poor, on the other hand, were not allowed to leave, as they were seen as the carriers of the disease, and could infect others.

The famous London diarist Samuel Pepys kept a diary throughout the Great Plague. In it, he recorded the main events in London in the scorching hot summer of 1665.

> 7th June
>
> The hottest day that I have ever felt in my life. In Drury Lane I saw two or three houses were marked with a red cross and the words "Lord have mercy upon us" written on the door. This worried me so much that I bought a roll of tobacco to smell and chew.

At the time of the Great Plague, nobody knew the cause of the infection. Some people believed that poisonous air spread the disease, while others blamed cats and dogs—as a result, thousands were destroyed. On August 31, Pepys entered the following into his diary:

> Fires were lit to keep the air clean. . . . All dogs and cats are to be caught and killed. People are being paid to kill them. Even public entertainment has been stopped, there's no theater, no sports and no games. Any houses containing plague victims have to be sealed for 40 days.

The Great Plague claimed around 100,000 victims, but by wintertime it had gone.

Experts believe that the Great Plague was caused by a bacterium called *Yersinia pestis*, which lived inside fleas. These fleas were carried by black rats, and when the flea bit a person, it would transfer the bacteria, resulting in the victim suffering from the plague. Historians think that plague-carrying rats and fleas were brought to England on boats from Europe. However, some experts take a different view. They argue that because the Black Death spread so quickly, it must have been caused by a virus that could be passed from person to person, like the disease Ebola, which also causes massive internal bleeding.

where can you find the "mound of ears"?

Toyotomi Hideyoshi (1536–98) was the ruler of Japan in the late 1500s. He was a son of a peasant, and began as a foot soldier in service to Oda Nobunaga. His exceptional military talents ensured his quick climb through the ranks to become Nobunaga's premier general. However, Nobunaga was assassinated in 1582, and so Hideyoshi became the ruler of Japan. In the 1500s, Japan was made up of principalities that were almost constantly at war with one another. Both Nobunaga and Hideyoshi helped to unify the country. Hideyoshi had visions of conquering China via Korea, and in the 1590s, Hideyoshi and his troops invaded Korea.

There, Hideyoshi and his men hacked off tens of thousands of Koreans' heads, including those of women and children. The taking of heads in battle was an age-old practice in East Asia. The victorious armies would take the severed heads of their defeated enemy and present them to their commander, as a proof of victory. Hideyoshi wanted to ship the heads back to Japan. However, it proved too impractical, as the heads took up too much room—and no doubt were extremely smelly and unpleasant—so instead he decided to just send back the ears and noses in barrels of salt water to preserve them. The trophies were systematically recorded against each individual warrior's name and shipped back to Japan, where a mound of them was created, which is now called the "Mound of Ears" (Mimizuka). The 30-foot-high (9-m) war memorial can still be seen in Kyoto.

what was the unpleasant job of a "decrotteur"?

In the 1700s, Paris was the world's cultural hub: the center of science, art, fashion, and generally a city of exquisite taste. Nonetheless, it was also, according to Louis-Sébastien Mercier, a French playwright of the time, "the center of stench." In a memorable account, Mercier described a main street in Paris as being "covered with stagnant water and dead cats."

According to Jean-Jacques Rousseau, the eighteenth-century French philosopher, in the courthouse, in the museum, at the opera, even at the palace of Versailles, "one does not know where to sit in summer, without inhaling the odor of stagnant urine." Courtyards, corridors, and alleys would be full of urine and feces. In addition, urine and feces were thrown from chamber pots being emptied out of windows, which often landed onto the heads of people passing by; the person throwing their excreta would call out, "*Garde l'eau!*" which meant literally, "Look out for the water!" Over time, the streets became choked with excrement. Carriages tearing down the streets would throw up the sludge, splashing pedestrians with rubbish and raw sewage.

So what was a person to do if attending some sophisticated event, at which an outfit matted with excrement might be regarded as somewhat outré? Luckily, Parisian high society catered for just this problem, as the grand houses would employ a person called a "decrotteur," which literally meant a "de-shitter." It was the decrotteur's rather unglamorous responsibility to wait at the door, and to clean the shoes and stockings of new arrivals so that they might enter polite society in as respectable and dignified a manner as possible.

in which society do men tie their penises into knots?

Warriors of the Karamojong tribe in northern Uganda believe that possession of a very long penis is a wondrous thing. To achieve a suitably impressive member, a warnor ties weights to the end of his penis, a process that begins when a boy reaches puberty. Over time, the amount of weight attached is steadily increased, which can result in a penis reaching as much as 18 inches (46 cm) in length. However, all this process achieves is to stretch the penis, so although it does become longer, there is a commensurate decrease in girth, so the penis becomes extremely thin. And since a very long, thin penis can get in the way of everyday chores, the tribesmen will often tie them into a knot or two when not in use.

In South African Bantu groups, elongated labia are regarded as a sign of great beauty, so the women of this society will go to great lengths to ensure, well, a great length! They achieve this by regular tugging on the labia minora. There is a similar tradition in southern Tanzania, where the traditional Maasai women elongate their labia through massage.

where might you see a penis 5 feet 6 inches (1.5 m) long?

The Icelandic Phallological Museum was founded by teacher and writer Sigurður Hjartarson, and boasts more than 260 penises. It began as Hjartarson's hobby in 1974. He explains his strange hobby by simply stating, "I wasn't interested in collecting stamps," so when someone kindly gave him a bull's penis many years ago, which apparently resembled a riding crop, he decided that he would become a collector of penises.

Since then he hasn't looked back, and his museum proudly displays penises of all shapes and sizes, ranging from the enormous 5 feet 6 inch (1.7 m) specimen of a sperm whale, to a 0.08 inch (2 mm) hamster penis bone, which really needs to be observed with the help of a magnifying glass. Most of the museum's penises have been donated, but Hjartarson has had to pay for some exhibits, including the elephant penis, which is nearly 3 feet 3 inches (1 m) long and is stuffed and mounted on a wooden board. Other exhibits include the penises of skunks, wallabies, and bears, which are, depending on their size, displayed either in special tanks or in humble glass jars. Surprisingly perhaps, the museum at present does not include in its collection a human penis, but Hjartarson is confident that this shortcoming will soon be rectified, as cocksure men from all parts of the world have generously volunteered to donate their penises after their death.

how does disney's snow white differ from the original gruesome tale?

Walt Disney's *Snow White and the Seven Dwarfs* (1937) is an enchanting tale about a beautiful young princess who is forced to work in rags as a maid for her jealous stepmother, the wicked queen. When a young prince shows interest in Snow White, the queen commands one of her huntsmen to kill the beautiful princess. However, as the huntsman is fond of Snow White, he doesn't kill her, and she is consequently looked after by seven dwarfs. After being poisoned by the queen, Snow White is reawakened by the kiss of a prince.

However, in the original story by the Brothers

Grimm, there's quite a bit more gore. The Brothers Grimm were two German brothers, Jacob and Wilhelm, who were famous for collecting folk tales, such as *Sleeping Beauty*, *Snow White*, and *Hansel and Gretel*, and publishing them in the 1800s.

In the Grimms' *Snow White*, the queen orders the huntsman to bring her the heart and the tongue of Snow White. The huntsman takes Snow White to the forest, draws his knife, and when he is about to pierce her heart, she begins to weep and begs for her life. The huntsman kills a boar instead and gives the queen the boar's heart and tongue, which she eats, believing them to belong to Snow White. At the end, when Snow White is revived and marries the prince, the evil queen attends the wedding. She is surprised when a pair of metal shoes made of iron are brought out, flaming hot from the oven. The wicked queen is forced to put them on, and has to dance until she falls down dead.

who carried around sir walter raleigh's head for nearly thirty years?

It is well known that the celebrated English adventurer Sir Walter Raleigh (1554–1618) was a key figure in popularizing tobacco smoking in Britain. During the 1500s, tobacco was seldom used for pleasure but was seen as a "cure-all" to treat almost every illness imaginable. Everything from skin complaints to gangrenous limbs was treated with tobacco, which was mostly taken as snuff. After taking up pipe-smoking, Raleigh began to promote and sell tobacco to the rich. The upper classes embraced the habit of pipe-smoking, and soon tobacco could be purchased in pubs and shops. Its popularity made Raleigh a lot of money.

James VI of Scotland not only disliked smoking but also hated Raleigh. When he became King of England in 1603, following the death of Elizabeth I, he had Raleigh, who was a Protestant, arrested and tried as a conspirator in a planned Catholic uprising. Raleigh was found guilty, even though the evidence was flimsy at best, and was later sentenced to death.

Raleigh was beheaded in 1618 at the age of sixty-four, and his body was buried in St. Margaret's Church, next to Westminster Abbey. However, his head was put into a red leather bag and given to his widow, Elizabeth. She had Raleigh's head embalmed and is said to have carried it around with her wherever she went. Apparently, when people came to visit her at home, she would ask them if they'd like to see Walter. If they agreed, she would bring it out of the bag. She continued carrying the head around right up to her death twenty-nine years later. But that was not the end of the story, because, even after Elizabeth's death, her son Carew continued the tradition, carrying his father's head with him everywhere he went. When Carew eventually died in 1666, Raleigh's father's head was buried with him, in West Horsley, Surrey.

chapter four

pernicious practices

for what unusual purpose did hippocrates recommend the use of pigeon droppings?

Hippocrates was an ancient Greek physician who is widely regarded as the "father of medicine" and the founder of a more modern, scientific approach to illness that rejected superstition and divine influence. One of Hippocrates's more unusual recommendations was that bald men should apply pigeon droppings to their bald spots, to encourage the hair to regrow. This may sound bizarre, but it is just one of many crazy cures for baldness that have been suggested since the earliest times, as throughout history bald men have tried just about anything to stimulate hair growth. The ancient Egyptians, for example, recommended the application of rotten crocodile or hippo fat, thinking that the bad smell would stimulate growth.

Ancient Romans had an ingenious use for houseflies. They thought that if they mashed them up into a paste, which was then applied onto a bald spot, the hair would grow back. Julius Caesar was said to have been follicularly challenged, so Cleopatra applied a substance containing ground horse teeth and deer marrow to his bald spots.

And the wacky remedies didn't die out with the

Ancient world. Even as recently as Victorian times, desperate slapheads were applying chicken feces to their bald spots and encouraging cows to lick their heads!

who has the world's longest ear hair?

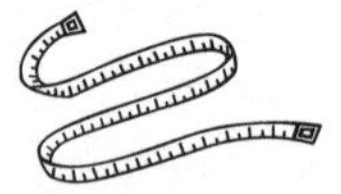

According to *The Guinness Book of World Records*, the person with the world's longest ear hair is Anthony Victor, who lives in India. At its longest point, Mr. Victor's ear hair measures an incredible 7.12 inches (18.1 cm).

Other revolting and unusual records include:

The longest leg hair belongs to Wesley Pemberton, who lives in the United States; it measures 6.6 inches (16.8 cm) long. It was measured on the set of *Lo Show dei Record* in Madrid, Spain, in 2008.

The record for the world's loudest burp belongs to Paul Hunn, who lives in the UK. Hunn set his record in 2008 by letting out an ear-splitting burp of 107.1 decibels, equivalent to a jet engine at take-off.

Kim Goodman of the United States can pop her eyeballs to a protrusion of 0.47 inches (12 mm) beyond her eye sockets. This astonishing feat was measured in Istanbul, Turkey, in 2007.

In 2001, Ken Edwards, who lives in the UK, set a record in London by eating thirty-six cockroaches in one minute. He said eating them was like "having an anesthetic at the back of the throat."

The world's longest tongue belongs to Stephen Taylor of the UK, whose prominent protuberance measures 3.74 inches (9.5 cm) from its tip to the center of his closed top lip. It was measured on the set of *Lo Show dei Record* in Milan, Italy, in 2006.

Until recently, the longest fingernails in the world belonged to a lady named Lee Redmond of the United States, who had not cut her nails since 1979. She had a total nail length of 28 feet 4.5 inches (8.65 m), as measured on the set of *Lo Show dei Record* in Madrid, Spain, in 2008. The longest nail was the right thumb, which was 2 feet 9 inches (90 cm) long. However, in February 2009, Redmond was involved in a car crash, in which her nails were damaged beyond repair.

According to *The Illustrated Book of Sexual Records*, one woman had pubic hair that reached down to her knees. The remarkably hirsute Maoni Vi of Cape Town, South Africa, had pubic hair that measured 28 inches (71 cm) from her groin, while the hair from her armpits measured an amazing 32 inches (81 cm).

why did howard hughes store urine in jars?

Billionaire Howard Hughes (1905–76) inherited a fortune from his father, who was an inventor and manufacturer of oil-mining equipment and tools in Houston, Texas. By the time Howard was eighteen, both his parents had died, leaving him devastated. At twenty-one, he became the legal holder of all the Hughes businesses.

However, Howard was not content to simply carry on the family business. Instead, he set his heart on success in Hollywood. In 1927, he produced his first movie, *Everybody's Acting*, which was followed the next year by *Two Arabian Knights*. Both were financial successes, with the latter film also winning the Academy Award for Best Director of a Comedy Picture. As Hughes went from success to success in Hollywood, he was simultaneously making a name for himself as a pioneering aviator. Hughes set multiple world air-speed records; in 1938, he completed a flight around the world in just three days and nineteen hours. He was also a notorious playboy, and was linked with many of the most desirable women of the age, including Ava Gardner, Katharine Hepburn, and Bette Davis.

Today Howard Hughes is mostly remembered for his bizarre behavior in his later life. Hughes developed an incapacitating obsessive-compulsive disorder (OCD) and lived in fear of germs. One of his aides, Bob Roberts, described "The Old Man" sitting in a chair, staring at the wall. Hughes was naked except for a napkin spread over his groin. His long hair was white and dirty, and his toenails and fingernails were some 6 inches (15 cm) in length. Hughes's doctor wasn't allowed to touch him or speak to him, so they communicated by writing and exchanging notes. Reluctance to part with his feces caused Hughes to suffer with constipation, which was probably the cause of his bad case of hemorrhoids, and would have been particularly painful when wiping his bottom, considering his 6-inch-long fingernails.

His obsession with hidden germs that could potentially cause him harm meant that he avoided contact with possible sources of dirt and was constantly washing his hands. He also kept his curtains permanently closed and his windows and doors sealed with masking tape. He was concerned about germs in his carpet, so he walked around with tissue boxes on his feet. Hughes also insisted that his urine should not be flushed down the toilet but stored in jars instead, and he also stopped taking baths.

For the last fifteen years or so of his life, Hughes

never left his bedroom. Many hours were spent watching movies while lying naked on dirty sheets. Ironically, by lying in filth, believing that the outside world was too dirty, Hughes greatly increased his chances of catching the hepatitis he was so anxious to avoid.

could cat ear mites live inside a human ear?

In a series of gross and bizarre experiments, a curious New York veterinarian named Dr. Robert Lopez attempted to determine whether *Otodectes cynotis*, an ear mite usually found in cats and dogs, could also infest human beings. Lopez later described the study in an article titled "Of Mites and Man" for the 1993 *Journal of the American Veterinary Medical Association.* Lopez explained how the study came about: "A client, accompanied by her three-year-old daughter, brought in two cats with severe ear mite infestations. In the examining room, the daughter happened to complain of an itching chest and abdomen. The mother stated that the daughter frequently held the cats for long periods, like dolls. . . . After the cat's ear mite infestation was cleared up, I learned that the daughter's itching quickly disappeared. At that time (1968), a search of the literature did not reveal any report of *Otodectes cynotis* infestation in human beings so I decided to be a guinea pig."

He began his experiment by transferring ear mites from a cat. First, he moistened a sterile cotton-tipped swab and transferred approximately 1 gram of ear mites from the cat to his left ear. Immediately, he began to hear scratching sounds, and then moving sounds, as the mites began to explore the inside of his ear canal. Then the area occupied by the mites began to itch, followed by noise and pain that intensified.

As the day went on, the itching increased and the sounds in his ear became louder as the mites traveled deeper toward his eardrum. For the next five hours, the mites were very active, but then their activity seemed to level off. At about 11 p.m., Lopez went to bed, at which point the mite activity began to increase again, and by midnight, they were busy biting, scratching, and moving about. By 1 a.m., they were making a lot of noise, and about two hours later, the itching and scratching was very intense, making sleep almost impossible.

Then, suddenly, the mites seemed to lessen their activities, the noise and itching decreased, and a brief sleep became possible. However, it seems that Lopez wasn't the only one sleeping, as the next morning the mite activity began again at 7 a.m. This pattern of mite activity was repeated the next day, and continued. By the third week, the ear was filling up with debris, and the hearing in Lopez's left ear had gone. By the fourth week, Lopez estimated that the mite activity had dropped by 75 percent, and he could feel the mites crawling across his face at night. The mites didn't try to enter his right ear, nor did they bite or cause any itching anywhere else on his body. At the end of one month, he could no longer hear or feel any mite activity, and the itching and noises were lessening. He decided to clean his ear for the first time since starting the experiment, and by the sixth week, there was no itching and hearing was normal.

Lopez carried out this experiment twice more on his left ear and, interestingly, he found the last experiment to be less itchy and less painful than the previous ones, and by the end of day nine, the mites had stopped biting altogether. He concluded that perhaps mammals have an immune response to these ear mites, which is why subsequent infestations are less intense.

In 1994, Lopez was awarded the Ig Nobel Prize in the field of entomology—the Ig Nobel Prizes are a parody of the Nobel Prizes and are designed to highlight amusing and thought-provoking research. At the ceremony, Lopez recited an original poem about ear mites, and then handed a selection of dead bugs to members of the audience.

why did the chinese bind the feet of young girls?

There is an old Chinese saying that states, "If you care for a son, you don't go easy on his studies; if you care for a daughter, you don't go easy on her foot-binding." Until fairly recently, the Chinese had a long tradition of binding the feet of young girls. Small feet were highly prized as a mark of great beauty; since women tend to have smaller feet than men, it was felt that the smaller a woman's feet, the more feminine she was.

Foot binding went on for almost a thousand years and only began to die out at the beginning of the twentieth century. Foot binding was prohibited in China in 1911, but continued in isolated regions well into the 1930s. Young girls would have their feet bound from the age of about seven. Foot binding would be carried out by the girl's mother and other close female relatives, and involved tightly wrapping each foot in bandages, sparing only the big toe, which pointed upward. The bindee would then be made to walk and put her weight onto the newly bound foot. The brutal object of this process was to break many of the bones in the foot, to bend the toes into the sole of the foot, and to bring the sole and heel as close together as possible, which sometimes caused toes to fall off. This was an excruciatingly painful process, and the girls would suffer pain for about a year. Eventually, however, their feet simply became numb. Every few weeks, the bound feet would be squeezed into smaller and smaller shoes. The goal was for the girl to eventually have a foot that measured as little as 3.5 inches (9 cm) long.

Naturally, the girls' feet would become grossly deformed, which meant they would hobble when they walked, if they could walk at all—many had to be carried. One man described his sister's foot-binding experience: "Auntie dragged her hobbling along, to keep the blood circulating. Sister wept throughout but Mother and Auntie didn't pity her in the slightest, saying that if one loved a daughter, one could not love her feet."

after the execution of king charles I, why did people dab their hankies into his blood?

Charles I (1600–49) was a deeply unpopular king who engaged in a disastrous power struggle with Parliament, which resulted in the English Civil War. In the First Civil War (1642–46), Charles was defeated, and he was then expected to accept

Parliament's demands for a constitutional monarchy. Instead, he escaped to the Isle of Wight and attempted to forge an alliance with Scotland. This led to a Second Civil War (1648–49), in which Charles was again defeated. This time, he was captured, tried, and found guilty of making war on his own people. He was sentenced to be executed in 1649. Charles was dignified and calm in the face of death, and on the scaffold he said, "All the world knows I never did begin a war with the two Houses of Parliament. . . . They began these unhappy troubles, not I. . . ."

As the ax came down upon his neck, there was a groan from the crowd. The axman then held Charles's head up to the crowd, and blood splattered everywhere. After paying a fee, people hurriedly dipped their handkerchiefs in the king's blood and took hairs from his head and beard. They also scraped up bloody earth from beneath the scaffold. People believed that royalty had godlike powers and thought that the king's blood would cure their wounds and illnesses. They wiped their blood-soaked hankies onto wounds, and some people dabbed it over themselves, one another, and their sick children.

After the execution, Charles's head was sewn back onto his body, and he was buried at Windsor Castle. However, in 1813, Royal Surgeon Sir Henry Halford was asked to perform an autopsy on Charles's body, and subsequently stole a vertebra from the corpse, which he took home and used as a saltcellar. Apparently, he enjoyed using it to shock his guests at dinner parties. Halford kept the bone for thirty years. However, when Queen Victoria found out, she was not amused and ordered that it be returned to Charles's skeleton.

why did many gladiators commit suicide?

During the heyday of the Roman Empire, the state would often provide lavish entertainment on public holidays, which might consist of theatrical performances, wild animal shows or gladiatorial contests.

Hundreds of thousands of people were sent to their deaths in the arenas of ancient Rome, through either battling other gladiators, or being forced to fight wild animals. Most gladiators were not willing participants in these contests; most were forced to fight, having been drawn from the ranks of slaves, criminals, or prisoners of war (although, strangely enough, there were some volunteers). New recruits received harsh training in combat skills, using wooden weapons, in preparation for the contest. Helmets, shields, and arm defenses were issued to fighters both in training and on the day of the contest.

The vast majority of gladiators were men, but for the novelty factor, occasionally women would be forced to fight. There are even reports of female gladiators fighting with dwarves, which apparently proved very popular. The presence of female gladiators in the arena steadily increased over the years, especially during the reign of Nero, in the first century A.D. Eventually the use of female gladiators was outlawed in the early third century A.D.

During the buildup to a gladiatorial contest, one bizarre ritual was that on the night before the fight, all the gladiators—who would spend the next day attempting to murder one another—would sit down and eat dinner together. Fans could pay to come and watch these dinners.

On the big day, the gladiator would be expected to fight against wild beasts, such as bears and lions, or against other unwilling participants. If lucky, gladiators could reasonably expect to fight only two or three times before being killed in the arena. Fearful of having an incredibly painful death, many gladiators tried to either escape or commit suicide. Authorities would keep constant watch over the gladiators, and shackle them with chains, to prevent them from escaping or committing suicide before the contest. Nonetheless, some still managed it.

On one famous occasion, a wealthy politician called Symmachus ordered a number of prisoners to fight in honor of his son. Twenty-nine Germans, who were supposed to fight the next day, instead decided to strangle one another. An even more gross method of suicide was employed by another German, who found a way to kill himself when he was allowed to go to the bathroom. Romans did not use toilet paper; instead, each bathroom would be equipped with a sponge on a stick. The German gladiator, desperate to avoid a terrifying and painful death in the arena, took one of these sponges and forced it down his own throat, suffocating himself.

what is a "plate woman"?

Some cultures practice a range of extreme and unusual methods to make themselves appear more attractive. Among the Mursi and Surma people of southwest Ethiopia, the adult females are known as "plate women." When they are in their early twenties, a small plate, called a "labret," is inserted into a slit separating their lower lip and jaw. After some time has passed and the flesh has stretched, the plate is replaced with a larger one. This process is repeated until the upper lip protrudes as much as 4 inches (10 cm), and the lower lip extends as far as the size of a dinner plate. However, the technique varies from tribe to tribe; some stretch only one lip, some both lips.

Lip plates are worn by women as a form of decoration and a sign of status. The plates must always be worn in the company of men, but when the women are alone, eating, sleeping, or in the company of other women, they are allowed to remove them. When the plate is removed, the women's stretched lips hang down loosely, swaying below their jaws. Ironically, it seems that this bizarre practice started as a way to make the women unattractive to slave traders, in the hope that they would go elsewhere in their search for slave girls, but eventually over time it became regarded as a sign of beauty.

Altering the teeth for aesthetic reasons is another peculiar beauty practice that has been found in cultures around the world. For example, some Australian Aboriginals will chip either one or both of their front incisors, to make them look more attractive. In parts of Africa and Central America, teeth are filed or chipped into points. In Bali, teeth are filed as an important rite of passage for adolescents. Mayans also filed and drilled holes into their teeth so that they could insert jewels.

In parts of Myanmar, it is seen as very beautiful for women to have unusually long necks. The women of the Padaung tribe are thus required by local custom to start wearing brass neck rings around the age of five. At first, they begin with five brass rings around their necks, but over time this number is steadily increased, and the ladies' necks can reach heights of around 15 inches (38 cm). The neck muscles become so stretched by this practice that it is believed that if an average tribeswoman were to remove the heavy brass rings, her neck would be unable to support her head.

what is urine therapy massage?

One treatment you might not see advertised at your local beauty salon is an Ayurvedic treatment called urine therapy massage. This treatment uses urine, either cold or heated, to massage the body. It is said that "the best urine to use is the first flow in the morning; about 8 to 10 ounces is most effective" (S.V. Govindan, *Ayurvedic Massage for Health and Healing*, 2000). The urine is then boiled and allowed to cool before being poured into a bowl and used to carry out a body massage. After the massage, the patient is advised not to take a shower or bath for at least two hours.

For centuries people have been availing themselves of urine's supposed curative powers by regularly drinking it, and urine therapy has been regaining popularity in recent years. Also, the fact remains that everybody has drunk their own urine at some point in their life, namely in the mother's womb—as babies urinate, their urine is released into the amniotic sac and becomes part of the amniotic fluid. Indeed, urine therapy is by no means new. The *Ebers Papyrus* is an Egyptian book on medicine, written in about 1000 B.C., which contains fifty-five recipes for urine therapy. Even the father of medicine, Hippocrates (460–377 B.C.), was an advocate of the practice of uropoty, which means the drinking of urine.

Religious use of urine therapy is well documented in India. The scripts of the ancient Sanskrit text *Damar Tantra*, which date back five thousand years, advise that drinking urine can rejuvenate the body. Animal urine was also believed to be therapeutic and, in early times, Hindus would accompany their cows to graze in fields, and would drink a cow's first urine of the day. Cow's urine was believed, and in fact still is by many Hindus, to have medicinal value.

in which society do the men claim never to fart?

The men of the Chagga tribe in Tanzania undergo a strange initiation ritual. At a given point during puberty, the boys of this tribe are taken away to take part in a ceremony in which their anuses are said to be blocked with plugs, so that from this point on, those initiated never again need to fart or defecate. Once the ceremony has taken place, the boy in question is required never again to admit to any kind of anal emission, especially in the presence of women, young girls, or uncircumcised boys.

On the day of initiation, the boy is taken from his home and is later returned to the tribe with his bum bleeding. The women are told that his bottom has been sealed, and that from that day forth, the women and children are expected to believe that he will never again fart or do a poo. In actual fact, of course, no operation is ever performed. Instead, all that happens is the boy is forced to sit repeatedly on a bush of thorns, and it's this which causes the bleeding. During the ceremony, the boy is ordered never again to pass wind, or anything else, except in secrecy, where no woman might find out about it. And then, when they eventually reach old age, the men of the Chagga tribe are again taken away so that the "operation" can be reversed. After this point, they can once again openly fart and defecate to their hearts' content.

In fact, the Chagga's view of masculinity seems to be bizarre in a number of respects. When, for instance, two men of the tribe become blood brothers, they seal the bond by drinking beer that is mixed with saliva and blood.

why did turkish wet nurses used to suckle puppies?

Although it sounds pretty gross to us, the idea of women suckling animals has been reasonably widespread in many cultures down the years. Various animals, and puppies in particular, have been used by nursing women for centuries for a number of different reasons. For example, Turkish wet nurses used to bring puppies along to their place of employment, as suckling the puppies would ensure that they kept producing milk. In Persia and Turkey, women would put puppies to their breasts when the nipples were not sticking out properly for breast-feeding.

The first American pediatric author, Dr. William Dewees (1768–1841), wrote a guide called the "Treatise on the Physical and Medical Treatment of Children," in which he suggested that pregnant women in the eighth month of pregnancy should allow "a young but sufficiently strong puppy" to suckle at their breasts, so as to "harden and confirm the nipples, improve breast secretion and prevent inflammation of the breast."

E. G. Rogers in *Early Folk Medical Practices in Tennessee* (1941) described the actual use of a puppy or a small pig. He claimed that one woman had nursed a baby pig for four weeks, and after that "the little pig followed her in and out of the house squealing."

As if this wasn't yucky enough, there is a corresponding tradition of children being breast-fed by animals. In France and Germany in the 1500s, many children were suckled by goats, sheep, and other animals. If a mother or a wet nurse was not available to breast-feed a baby, the next best thing was for them to suckle at the teats of animals. In 1580, the French author Michel de Montaigne (1533–92) wrote, “It is common to see women of the village, when they cannot feed the children at their breast, call the goats to their rescue.”

what is hara-kiri?

From around A.D. 700 to the 1860s, shoguns (military dictators) ruled Japan, and Japanese warriors called “samurai” served the shoguns. Samurai warriors were expert fighters and carried two swords. Sword makers would test their swords’ effectiveness on either corpses, or condemned prisoners. They would carry out a series of tests by cutting through body parts, starting with small bones, such as those of a finger or a toe, and then moving on to bigger bones, such as an arm or a thigh bone.

The samurai believed that if you got beaten in battle, the noble thing to do was to commit suicide. They did this by slicing open their abdomen, a gross practice known as “hara-kiri,” or belly-cutting. After committing hara-kiri, the warrior would have his head cut off. Individuals also committed hara-kiri if they did something that was shameful or had dishonored their family or associates. In fact, the offended people were invited to witness the death.

A Japanese family called the Hojo clan ruled Japan from about 1185 until 1333. In 1333, the city of Kamakura was invaded and taken over. When the battle was seen to be lost, the Hojo regents were determined to die like true samurai. In total, 870 samurai and their families committed suicide. One samurai "cut his body with a long cut from left to right and fell down, pulling out his intestines . . ." Other samurai cut open their abdomens too, while some were reported to have cut off their own heads.

The samurai showed they were great warriors by collecting the heads of their victims. They chopped off the heads of other samurai warriors, put them into a head bag, and got their wives to prepare the trophies for display. This preparation would include applying makeup, dressing the hair, and blackening the teeth. Black teeth were considered very fashionable at that time. The samurai would then impress their warrior friends by showing off their head collections. The better the heads looked, and the sweeter they smelled, the greater the victory; a perfumed head meant they had killed a top samurai.

In 1213, a battle between samurai warriors resulted in 234 heads being placed along the banks of the Katasegawa River.

what people tickled their throats to induce vomiting?

Roman diners regularly spat and vomited at the table. In fact, Romans were known to tickle their throats with feathers after each meal to induce vomiting, thus allowing them to return to their gluttonous feasting. They even had special bowls in which to spew, although they frequently decided to spit or puke on the floor instead. Slaves had the disgusting job of cleaning up the mess. The puke would have consisted of delicacies such as dormice drizzled with honey, which was a favorite food of the Romans. The philosopher Seneca (4 B.C.–A.D. 65) wrote, "The Romans vomit that they may eat, and eat that they may vomit." It's amazing anyone had an appetite; you'd think the smell on its own would turn stomachs.

In the Middle Ages, it was considered acceptable to spit during dinner but only onto the floor. It was also quite proper to belch at the table but not in somebody's face. Picking one's nose was also fine, providing the snot was wiped onto one's clothing or the tablecloth. Guides discussing table manners were written during this time, which observed that spitting was not only a custom but that people clearly felt the need to spit frequently. However, although spitting was commonplace, an individual was not allowed to spit on or over the table, but he could spit under it. Nor should a person spit into the washbasin when cleaning his mouth or hands, but he could spit beside it. In the 1500s, people who spat on the ground were then expected to tread on it, especially if it contained purulence (i.e., pus). However, by the late 1600s, it was regarded as indecent to spit in general.

how clean is your toothbrush?

It would seem reasonable to assume that your toothbrush is fairly clean, wouldn't it? After all, it gets rinsed twice a day, and one might presume that toothpaste kills most germs. In fact, however, chances are that your toothbrush is full of bacteria and fungi.

It may surprise you to know that bacteria can thrive on toothbrushes, as they provide them with lots of food and water. In one Australian study, the two most commonly found germs on used toothbrushes were bacteria called "staphylococci" and "streptococci." Researchers have also found that the viruses for influenza and herpes simplex—and the common fungus *Candida albicans*, which can cause mouth sores and vaginitis—can survive on toothbrushes. It is also not a good idea to use somebody else's used toothbrush, as research has shown that infection can be caught from it and lead to sickness. Even if you buy a brand-new toothbrush, it becomes contaminated the first time you use it, and bacteria and viruses from one brush can easily spread to another. However, it isn't actually clear how much of a danger this poses in the everyday life of a healthy person. And the good news is that most toothbrush germs disappear when the toothbrush has thoroughly dried out, so perhaps the answer is simply to rotate toothbrushes, so that they are regularly allowed to dry.

It also makes a big difference where you keep your toothbrush. A study in the journal *Applied Microbiology* showed that there is a risk of contamination if your toothbrush is positioned near the toilet. When we activate the flush, it causes water droplets to spray into the air, and it has been estimated to catapult water droplets containing more than 25,000 virus particles and 600,000 bacteria from the toilet. If the toothbrush is in this vicinity, it can become contaminated with airborne fecal bacteria. It has been found that these water droplets can travel as far as 20 feet (6 m) from the toilet. One solution is to put the lid down before flushing the toilet, but this will only mean the bacteria will instead be found on the underside of the toilet lid. Still, probably better that than having them on your toothbrush!

what causes the stench of b.o.?

Body odor is the unpleasant smell produced by a person's body if they don't wash often enough. Adults have more than 2 million sweat glands, each of which comprises a tube that opens at a pore on the skin's surface. Sweat moves up the tube and onto the skin, where it evaporates to help keep the body cool. There is also a special type of sweat gland called an "apocrine gland," which is found only under the arms and between the legs, and these glands produce a thicker liquid than the body's other sweat glands, and it's this thicker liquid that can give rise to body odor. Interestingly, sweat itself has no odor at all. But when sweat comes into contact with the bacteria on your skin, you can wind up with a terrible smell, because moist clothing is a great place for bacteria to reproduce.

Infrequent washing of the skin and hair can cause dirt, dead skin cells, and dried sweat to collect—particularly under the arms. This provides food for bacteria, which then release a foul odor. It's therefore these bacteria that cause the smell of body odor. Fungi can also grow on dirty skin, and these can cause illness. For example, a condition called "athlete's foot" is caused by a fungus that attacks the skin between the toes, and in more severe cases can spread over the whole foot. It causes the skin to peel off the foot, leaving itchy, sore patches. Fungi can also affect the nails, resulting in an unsightly discoloration, crumbling, and separation of the nail. As well as the risks from bacteria and fungi, dirty skin also makes an ideal breeding ground for lice, mites, and fleas, so the message is clear—scrub up!

how often did people wash in the 1600s?

From the Middle Ages onward, many people believed bathing was unhealthy and could lead to illness; it was even considered a sin. In the 1600s, people weren't too concerned about personal hygiene. If they bathed at all, the wealthiest people used a wooden or copper tub in front of the fire, in water that would have been heated in the kitchen and then brought upstairs in buckets by the servants.

Soap was very expensive, and so if people did wash, they would only use it on their faces, necks, hands, and occasionally their feet. Unsurprisingly, then, people of the time reeked of B.O. and had foul-smelling breath. To hide the pungent smell of B.O., women carried ornately embroidered bags, tied at the waist, containing aromatic potpourri. In fact, perfume was invented to cover up unpleasant body odors. At this time, cleaning the teeth was done by rubbing them with a piece of cloth and using toothpicks made from wood or ivory to pick out trapped bits of food.

Elizabeth I was considered something of a neat freak, as she was known to have a bath as often as once a month! However, Elizabethans were fastidious about some things, and many would clean their nails and comb their hair every day. Queen Isabella of Castile (1451–1504) was responsible for sponsoring Christopher Columbus's voyage of discovery to the New World. Legend has it that the queen boasted that she bathed only twice in her life—at birth and before her marriage.

which room contains the most germs—the bathroom or the kitchen?

Dr. Charles Gerba, known as "Dr. Germ" by his colleagues, is a microbiologist at the University of Arizona. Dr. Germ is renowned for being America's expert on domestic and public hygiene, and has appeared on numerous television programs. He has made quite a name for himself studying public toilets and other places where germs hang out.

Dr. Germ carried out a survey and found that in most homes, the bathroom is actually much cleaner than the kitchen. Surprisingly, the least contaminated place in either room was the toilet seat. Accordingly, Dr. Germ concluded that most people would be safer making a sandwich on top of the toilet bowl rather than in the kitchen. He also quipped, "That's why your dog likes to drink out of the toilet!"

Moist environments in the kitchen, such as the dishcloth and the kitchen sink drain, were some of the most contaminated with bacteria. However, the worst offender is the kitchen sponge, which can contain up to 50 million bacteria! Fecal bacteria was found in kitchen sponges and dishcloths. Dr. Germ found that kitchen sinks are the place where you can find the worst kinds of germs, and thorough wiping of sink and counters with a sponge serves only to spread them all over the kitchen. He also found that 20 percent of coffee cups contained fecal bacteria, which had almost certainly been transmitted from the sponges that had been used to wash them.

Ironically, it may be that the people who try hardest to keep their kitchen clean may be those in the most danger. Dr. Germ explains, "Clean people are usually the dirtiest, because they spread the germs all around. The bachelor who never cleans is usually the cleanest from a germ standpoint." If you want to actually clean your kitchen, rather than simply pushing the germs around, Dr. Germ recommends using an antibacterial spray and paper towels. However, most people who use their kitchens every day rarely get infected with anything, because most of the germs are fairly benign, and the body's immune system helps to fight off harmful germs.

where else are we likely to encounter germs?

There are an estimated 5,000,000,000,000,000,000,000,000,000,000 bacteria living on Earth—that's 5 million trillion trillion—and an astonishing number of these live on or around our office desks. According to a study conducted by the famous Dr. Germ, the average desk harbors four hundred times more bacteria than the average toilet seat.

However, the germiest thing in the workplace is actually the phone. An average phone receiver contains more than 25,000 germs per square inch, in comparison to an average toilet seat, which contains only about forty-nine germs per square inch.

The dedicated microbiologists of the University of Arizona have also sampled varied public places to determine the most germ-ridden areas. They found that surfaces that were frequently touched were often contaminated with bodily fluids such as blood, spit, mucus, and sweat. After you touch these surfaces, your hands then become contaminated by them, so that everything you touch then becomes contaminated. Then, for most of us, it's only a matter of time before our hands touch a part of our face, and if the nose is touched, germs can enter it and lead to infection. Research has shown that using soap and water to wash the hands reduces your risk of developing diarrhea by 45 percent and decreases your risk of catching other nasty bugs that cause intestinal infections by around 50 percent.

The most germ-ridden places were found to be playgrounds, nurseries, buses, shopping trolleys, escalator handrails, and public phones. Perhaps unsurprisingly, Dr. Germ is very cautious about which public surfaces he touches. "Personally," he recommends, "I use my knuckle—never my fingertip—to push the elevator button, especially to the ground floor, which is the most contaminated [button]." Dr. Germ also found that women's public toilets contained twice as much poo bacteria as men's, perhaps due to the fact that women are often accompanied by small children and babies whose nappies need changing.

Dr. Germ also recommends that if you wash your underwear, make it your last load. He says, "Basically, if you do undergarments in one load and handkerchiefs in the next, you're blowing your nose in what was in your underwear." Apparently, freshly washed laundry is heaving with bacteria, and a survey showed that half of all homes had fecal material in the washing machines, and as much as 0.02 pounds (10 g) of fecal matter was found in underwear. Washing didn't always help get rid of this, either, because most people do a washing machine load along with other clothes, thereby spreading around the fecal contamination.

Dr. Germ's study contained a number of other surprising findings. For example, an ATM was found to have more germs than a public toilet's door handle! One explanation for this may be that hands touching a public toilet's doorknob are more likely to have been washed immediately prior to touching it, compared to those hands that might touch an ATM keypad. One study even found fecal matter on the screen of an ATM.

The researchers used fluorescent dyes to track the germs volunteers touched during an average day, and to monitor how they spread them throughout their homes. When the household surfaces were exposed to black lights, the result was like Times Square on New Year's Eve.

how hygienic are we?

A survey of more than six thousand people was conducted as part of an exhibition titled "Grossology" at the Science Museum in London, and it revealed that the British have some pretty gross habits. A third of those asked about their personal habits said they picked their noses more than five times a day, and 34 percent admitted to eating their boogies. There were also regional variations, with the biggest nose-pickers being the Northern Irish, 44 percent of whom admitted to picking their noses more than five times a day. Thirty-nine percent of snot-loving Scots admitted to eating their boogies.

Many men and women admitted to not changing their underwear every day, including 39 percent of the Northern Irish. Overall, unsurprisingly, women changed their underwear more frequently than men. The men of Kent were particularly shameless, with only 3 percent of those questioned confessing to a daily change of pants, compared to 78 percent of Welsh men. Researchers also found that 34 percent of people had no shame in burping loudly in public, and that 29 percent would fart openly in public. This kind of earthy, primitive charm was particularly pronounced in Essex, where nearly 50 percent of men had no qualms about farting or belching loudly in public.

However, we Brits are certainly not alone in our mucky habits. In September 2000, the American Society for Microbiology reported the results of a survey that found that of 7,836 people observed using public toilets in five American cities—Chicago, Atlanta, New York, New Orleans, and San Francisco—only about two thirds washed their hands. The glamorous Gothamites of New York City were the least likely to wash their hands, with only 49 percent of them bothering.

what is the worst job in history?

The annals of history are full of disgusting, dangerous, and demeaning jobs, but one of the most unpleasant has to have been that of the "fuller." In the Middle Ages, this crummy job managed to combine grossness with mind-numbing banality. Fullers worked in the preparation of wool. After the sheep were sheared, a natural grease would remain in the wool. This grease made the cloth both coarse and widely meshed, so the job of the fuller was to get rid of the grease and other unwanted substances from the newly woven woolen cloth. To achieve this, the cloth was soaked in some kind of alkaline solution, and the cheapest alkaline solution available was stale urine, which is full of ammonia.

Of course, urine was always plentiful, and bucketfuls would be obtained from local houses, farms, and even public toilets. The next step was the tedious task of stamping on the cloth, while knee-deep in stale urine, for between seven and eight hours, to produce a finished piece of thick cloth. However, if insufficient attention had been paid throughout the treatment, it could lead to holes in the fabric and ruin the whole cloth, so the fuller would have to be very thorough. The fuller would then rinse the cloth in clean water, take it outside, and stretch it out to dry on hooks on a device called a “tenterframe.” This process of creating tension is the origin of the phrase “on tenterhooks.”

Another unpleasant profession was that of the “tanner” during the Victorian era. A tanner was an extremely skilled craftsman, but the job was very smelly. It was so smelly, in fact, that tanners were required by law to set up their operations on the outskirts of town, due to the noxious stink.

The tanning process would begin after the cow had been slaughtered and skinned. First, the tanner would soften the hide by soaking it in lime, then remove all the rotting meat and fat from it. The fat was useful, as it could later be made into soap. The cow’s hair would also be removed by scraping it off with a knife. It was important to remove all the hairs, because after the leather was tanned, the hairs would not come out. Then the leather would be soaked in a gross mixture called “bate,” which consisted of warmed-up water and feces, usually dog’s poo, to remove the lime and soften the hide. Later, the hide was soaked in tanning fluid for a year, cleaned off, and finally dried out so that leather items could be made from it.

If the tanner’s job sounds bad, spare a thought for his suppliers. Tanners needed a constant supply of dog poo, which was provided by somewhat lowlier workers called “pure collectors.” Their job was at least a simple one—they would roam the streets looking for dog poo, which they would scoop up, often with their bare hands, and then sell on to people such as tanners.

During the Middle Ages, the job of the "squire" involved looking after his knight's every need, including caring for the knight's armor and weapons, and helping him to train. The squire would also help the knight to dress, but the worst part of the job was undoubtedly cleaning the knight's armor after a heavy day on the battlefield. It took almost an hour to dress a knight in plate armor, which meant there were no quick toilet breaks while fighting. This meant that the knight not only sweated heavily, as tournaments mostly took place during the summer months, but also urinated and excreted inside the armor.

Therefore, the squire had the unenviable task of cleaning the armor, and to make matters worse, water was in short supply and far too precious to be used for trivial matters like cleaning, so the squire would instead have to carry out his distasteful chore using an abrasive mixture of sand, vinegar, and urine.

From the sixteenth century, it was the trade of the "saltpeter man" to extract potassium nitrates, also known as saltpeter, from feces and urine. Saltpeter was in considerable demand, as it was an important constituent of gunpowder. It was thus vital that the supply of saltpeter be maintained, and so to ensure this, the saltpeter men were given a license from the king to go into anyone's home and dig wherever they saw fit, which meant they had the right of access to people's houses, barns, stables, outhouses, cellars, latrines, pigpens and manure heaps. They would then dig out the excrement, and the sewage-ridden earth of the cesspit, and cart it away in barrels. How, you may wonder, would they know where to dig? Well, often their sense of smell would suffice, but where there was doubt, the saltpeter men would identify which parts of the soil were rich in saltpeter by tasting it.

Understandably, people weren't keen to have saltpeter men digging up their land, so to preserve their property from damage by the ruinous digging, landowners often resorted to bribery. Consequently, it later became an offense to bribe a saltpeter man. And if it wasn't enough that they would make a mess of your property, once the saltpeter men had dug up your earth, they then expected the landowner to supply cheap transport for it. In 1638, saltpeter men even tried to get permission to search for nitrate-rich substances in churches, arguing that "women pee in their seats, which causes excellent saltpeter" (Tony Robinson, *The Worst Jobs in History*, 2007).

In October 1660, Samuel Pepys complained in his diary: "Going down to my cellar . . . I put my feet into a great heap of turds, by which I find that Mr. Turner's house of office [cesspit] is full and comes into my cellar." The mess was cleared up five days later. Unfortunately, this was a common occurrence, as the cesspits were poorly constructed and often leaked.

what was a "dredgerman"?

Many of Charles Dickens's works make reference to London's River Thames. In his novel *Our Mutual Friend*, the river plays a central part and serves as a metaphor for the city's filth and corruption. At the beginning of the novel, two men are found drowned in the Thames by the dredgerman. The river is covered with slime, which mostly consists of sewage. As disgusting as the river was—Dickens's account was typically accurate—the Thames provided many poor Londoners with a living. In the late 1800s, the dredgerman's job involved fishing for corpses. In those days, dredgermen were kept busy, as the Thames was often littered with corpses. Dredgermen were paid for each corpse they found, but many of them would empty the pockets of the dead before handing the body over to the authorities.

Another similarly unpleasant job of the time was that of the "mud lark," although calling it a "job" may be something of a stretch. Mud larks were scavengers who spent their time trawling the banks of the Thames, looking for anything of value. In those days, it was common to see old women foraging along the riverbank, carrying a tatty basket or a tin kettle, into which they would put anything of value they might find. It was not just a job for women either; small children as young as seven years old would also be found searching through the filth. They made a living by paddling and groping for any sort of refuse washed up by the tide on the banks of the River Thames. They poked around in the wet mud for small pieces of coal, chips of wood, bits of old rope, bones, or more valuable debris such as copper nails.

In Victorian London, mud larks often had to wade up to their midriffs through excrement and mud while scavenging on the banks of the River Thames. They were too poor to buy decent clothing, so they wore filthy rags and didn't wear shoes. Consequently, it was common for them to injure themselves, or catch infection, by standing on nails or glass. But if a mud lark didn't find anything she could sell before the tide rose, she would most likely go hungry until the tide had next subsided, when she could once again wade through the grimy water.

are violin strings really made of "catgut"?

The strings of today's violins are almost always made of synthetic materials, but it is true that a few centuries ago, they were made from the intestines of animals. However, although this material is known as "catgut," it is in fact made from the intestines of sheep, not cats.

Violin string makers often worked at abattoirs, so that they could be near the source of violin strings, which was the lower intestines of sheep. To produce the strings, the string maker would slice open the sheep's abdomen, being careful not to damage the precious intestines, remove the intestine whole, and then knead the entire length of the gut, to squeeze out any poo that remained stuck inside. The skilled craftsman would then remove all fatty tissues, blood vessels, and bile, and clean the intestines thoroughly. Then the thicker parts of the intestine would be taken away to make sausage skins, while the thinner ends would be soaked in cold water, crushed, and then fumigated, by bathing them in a sulfur bath. Finally, the material would be twisted together and dried, resulting in the valuable and melodious violin strings. The whole job required a great level of skill and expertise but would also have been rather stinky.

what is "phossy jaw"?

In Victorian times, many poor girls were forced to work fourteen hours a day with few breaks, standing the whole time, for very little pay, in matchstick factories. The girls were even fined for offenses as trivial as talking, having dirty feet, or dropping matches. These "factories" were often little more than dingy, dark rooms with working conditions that routinely exposed the girls to a dangerous and unpleasant poison.

In those days, matches were produced simply by dipping small sticks of wood into a white toxic substance called phosphorus. Phosphorus led to a condition known as "phossy jaw," whose nasty symptoms included yellowing of the skin, hair loss, toothache, and the jaw swelling up with weeping abscesses. The entire jaw would then begin to rot, and the condition caused the girls to smell terrible, a smell that was described as being similar to a cross between stale urine and garlic. The only available treatment was removal of the jawbone, an agonizing operation that would disfigure the sufferer for life. From 1847 onward, there was a steady trickle of cases of phossy jaw in London hospitals. In the Bryant and May factory in London alone, there were fifty-one cases over twenty years. There was an eventual phasing out of white phosphorus matches in 1906.

why did the wari tribe of the amazon eat the bodies of their dead relatives?

Throughout history, different cultures have treated human remains in a variety of ways. In our culture, we currently consider it normal and respectful to burn them, or to freeze them, or even to drain them of blood, pump them full of chemicals, put makeup on them, and bury them in the ground. In the past, other societies, such as the ancient Egyptians, felt that bodies should be preserved for eternity through mummification. And in some primitive societies around the world, from Australia to South America, it was once considered a proper and fitting show of respect to eat the dead.

In the 1960s, a tribe living in the Amazon practiced two subtly distinct types of cannibalism. The Wari tribe would eat their defeated enemies, to show their anger and contempt for them. On the other hand, when their nearest and dearest died, the Wari believed it was important to demonstrate their love and respect, which they duly expressed by eating them. However, the Wari never ate their own blood relatives. This job was given to their in-laws, and in this society it would have been considered extremely rude not to oblige. The thought of their loved ones being buried and left to rot in the cold, wet ground was not considered dignified by the Wari tribe. An intact corpse was something painful to view, but if it was eaten, the mourners could deal with their loss more effectively. However, this practice has since ceased.

which famous criminals ate human body parts?

The "Butcher of Berlin" was the nickname given to Georg Karl Grossman, who was arrested in 1921 after having killed and chopped up several prostitutes. In particular, he had a preference for plump women. He was caught when neighbors heard him struggling with a woman who was fighting him off. Police burst in and found a woman's murdered body. Grossman gained his nickname because not only did he eat the flesh of his victims, he also sold a lot of meat, especially sausages, on the black market. It is believed that this meat was the flesh of his victims, which he sold in the form of sausages, at a cheap rate. He even had a hot dog stand at a railway station. It's estimated that Grossman killed about fifty women. At his trial, he showed no remorse and laughed when he was given the death penalty. Later, he hanged himself in jail.

The Russian serial killer Andrei Chikatilo, known as the Russian Ripper, didn't sell human flesh as sausage meat, but he did have a penchant for eating genitals. During the 1980s, he consumed the genitals of some fifty-three young women and children. Being a schoolteacher and Communist party member, Chikatilo was a respected and trusted member of the local community, which made it easier for him to commit his terrible crimes, as people who knew Chikatilo simply couldn't bring themselves to suspect him of murder. Under normal circumstances, Chikatilo was impotent; however, he found that he was sexually aroused by violence and cruelty, and could achieve orgasm by stabbing and raping his victims. Chikatilo said it gave him animal satisfaction "to chew or swallow nipples or testicles." After Chikatilo was caught, he confessed and gave a full account of his crimes. In 1994, he was executed by a single shot fired into the back of his head.

Another very disturbing recent case was that of a German computer technician called Armin Meiwes, who posted an advertisement on an Internet site that read, "Looking for a well-built 18- to 30-year-old to be slaughtered and then consumed." Bernd-Jürgen Brandes replied to the ad, agreeing to be eaten, and so the two men met up on March 9, 2001. On the day, they made a video. One part of this gruesome two-hour video shows Brandes urging Meiwes to slice off his penis, so that the pair can fry it and eat it. Meiwes does so, using a knife, and then cooks the penis in garlic. They both shared the penis meal but complained that, after cooking, it was too tough. Brandes joked, "If I'm still alive in the morning, we can eat my balls for breakfast."

Hours later, Brandes died of blood loss in Meiwes's bath, having been given large quantities of alcohol, painkillers, and sleeping pills, while Meiwes read a Star Trek novel. Over the next ten months, Meiwes ate about 44 pounds (20 kg) of his victim, which he stored in his freezer. Police also found a photograph showing Brandes's amputated foot, which was impaled with a fork and sat on a dinner plate covered with sauce. Meiwes is currently serving a life sentence in a German prison.

how many insect parts do we unintentionally eat?

Realistically, few of the people who read this book will ever deliberately sit down to a meal of fried hairy caterpillars or chocolate-covered cockroaches. However, even if we may not actively choose to eat insects, there seems to be little we can do to avoid them forming some part of our diet, as insect parts will crop up in our food whether we like it or not. In fact, those people who practice insect eating, or "entomophagy," to give it its proper name, have pointed out that although "only" 80 percent of the world's population eats insects intentionally, 100 percent of us do eat them, one way or another. According to estimates, the average person consumes about two pounds (1 kg) of insects every year. Harvard medical entomologist Dr. Richard Pollack has eaten numerous bugs, and he states:

> Virtually everything we eat has bugs (entire or parts) within; indeed there are government standards as to the maximum number of bug parts per unit for each type of food. Imagine the phenomenal biomass of insects and mites that get included when cereal crops are harvested and processed.

The U.S. Food and Drug Administration's (FDA) "Food Action Defect Levels" handbook itemizes the "levels of unavoidable defects in foods that present no health hazards for humans." The following is a list of foods and the amount of insect parts, rodent parts, and other general unpleasantness that is allowed to be contained within them.

Chocolate may contain up to sixty insect fragments per 100 grams, and an average of less than one rodent hair per 100 grams.

Macaroni and noodle products are allowed up to 225 insect fragments per 225 grams, and no more than four and a half rodent hairs per 225 grams.

Canned mushrooms should contain no more than five maggots of 0.07 inches (2 mm) or longer per 100 grams of drained mushrooms, and no more than twenty maggots of any size.

Peanut butter should contain fewer than thirty insect fragments per 100 grams, and an average of less than one rodent hair per 100 grams.

Canned tomatoes may contain up to ten fly eggs per 500 grams, and no more than two maggots per 500 grams.

Wheat should average less than 9 milligrams of rodent excreta pellets, and/or pellet fragments, per kilogram.

why have indian women been known to throw themselves onto funeral pyres?

The Hindu ritual of "sati" involves a Hindu wife following her husband to his death by ascending his funeral pyre with him, or ascending one of her own shortly afterward, to be burned to death. Sati originated more than seven hundred years ago, and was regarded as a tradition of the noble classes in India. As a widow, a woman loses her property and her social standing; however, she does not become a widow until her husband is cremated. Therefore, by performing sati, the woman is said to never lose her status as a married woman, and thus becomes divine.

By the 1800s, more than five hundred women a year were immolating themselves, until 1829, when the custom was finally outlawed. However, even since India gained independence in 1947, some forty cases have been reported, including that of a pretty, well-educated, eighteen-year-old girl named Roop Kanwar, in the town of Deorala in 1987. About eight months into their marriage, Roop's husband died of a burst appendix. According to reports, Roop, wearing her wedding sari, climbed onto her husband's pyre, took his head in her hands, and urged her brother-in-law to light the fire. He did so without hesitation, and she was engulfed in flames and burned to death.

However, while thousands of people were present at Roop's sati, there was little agreement about what had happened. Supporters of sati claimed that Roop had calmly seated herself on the pyre and willingly submitted to the flames, without once crying out. Opponents of the practice, on the other hand, said that Roop had tried several times to escape the flames, only to be pushed back by her in-laws. They argued that Roop would have been a financial liability to her husband's family, and so was drugged and coerced into submission.

A more recent case of sati occurred in the central state of Madhya Pradesh. In the summer of 2002, a sixty-five-year-old woman named Kuttu Bai was reported to have climbed onto the burning pyre of her dead husband, thereby committing the act of sati. She was watched by a supportive crowd of more than one thousand people that included her two sons, who made no effort to stop her. Reportedly, the villagers shouted slogans in support of the act. Policemen who attempted to stop the widow's self-immolation were beaten up and prevented from rescuing her.

According to reports, however, Kuttu Bai had not even had a particularly good relationship with her husband, or her eldest son, and there were many who believed she had been forced to her death. Fifteen people, including the woman's two grown-up sons, were arrested over the incident and faced charges of murder and conspiracy. Consequently, the sons and Kuttu's brothers were found guilty and sentenced to life imprisonment.

which well-known drug is made from horse's urine?

Before the introduction of Viagra, Premarin was the most widely prescribed drug in the United States. Premarin is a hormone replacement therapy that contains a combination of estrogens. When a woman goes through menopause, her level of estrogen falls by some 40–60 percent, which is the cause of many of the problems associated with menopause. Hormone replacement therapy can help a woman manage the symptoms of menopause, which can include hot flashes, night sweats, and vaginal dryness. Premarin comes in a number of different forms, including tablets, patches, and vaginal creams. It is also thought to help reduce the risk of osteoporosis and reduce the chances of heart disease in women over fifty.

Premarin and other similar products all contain estrogen derived from pregnant mares' urine. In fact, the brand name Premarin is derived from its origins—pregnant mares' urine. However, the use of Premarin has recently declined, probably because of widespread public concern about the cruel treatment of the horses from whom the drug is obtained. The manufacturers of Premarin keep pregnant mares in stalls and withhold water from them so that their urine is strong, as this makes the horse estrogen easier to extract.

what animal mates until he dies from exhaustion?

The male brown *Antechinus* is about the size of a mouse; has short brown hair, a pointed snout, and large ears; and belongs to the same family as the Tasmanian devil. It is widespread throughout forests in southeastern Australia, and feeds on a diet of beetles, spiders, and cockroaches. The *Antechinus* is notorious for its short and intense single mating season. At about eleven months old, the males begin a frantic two-week breeding frenzy, in order to father as many offspring as possible. Leading up to breeding season, males become increasingly aggressive toward one another as they compete for females. Males and females form pairs, and the male diligently guards her and fights off competitors who want to mate with his partner. Stress hormones in their bodies reach high levels, and this leads to reduced feeding and lowered immune systems but allows for sustained bonking. In this period, each male will mate for six to twelve hours at a time.

Unfortunately, after only one mating season, all this furious mating, along with the frequent aggressive clashes with male rivals, results in the animal's demise—he basically shags himself to death.

has anyone carried out a head transplant?

In the 1950s, Russian surgeon Vladimir Demikhov undertook an innovative experiment to create a two-headed dog. He grafted the head, shoulders, and front legs of a puppy onto the neck of an adult dog. Eventually, he created twenty of these hybrids, but most of them died shortly after the surgery because of infection. However, one of the two-headed animals survived for twenty-nine days.

In his book *Experimental Transplantation of Vital Organs*, Demikhov includes photographs of another experiment carried out in 1954: the transplantation of a one-month-old puppy's head and front legs to the neck of an adult dog. The notes imply that the transplanted puppy remained lively and content, but the adult dog would sometimes shake its head and body, to attempt to free itself of the unwanted appendage. At one point the adult dog bit the puppy behind the ear, causing the puppy to yelp and shake its head.

Another Demikhov experiment involved attaching a puppy to Pirat, a German shepherd. This time, the puppy was able to eat and drink on its own, though it didn't actually need to, because it received all its nutrients from Pirat. The puppy's foodpipe was not joined to a stomach—instead it emptied onto the outside of the dog—so when the puppy enthusiastically lapped at a bowl of milk, it dribbled out of the stump of its food pipe and onto Pirat's neck.

Later, Demikhov carried out an experiment that involved cutting dogs into halves. The upper part of one dog would then be attached to the lower part of another and their arteries grafted together. However, cutting through spinal nerves meant that the lower body would be paralyzed. Unsurprisingly, these experiments did not win much admiration from the mainstream scientific community.

chapter five

disgusting diseases, curious cures, and savage tortures

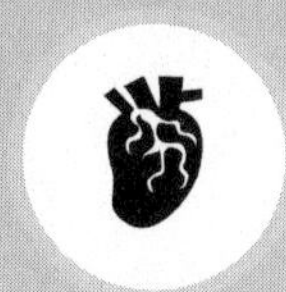

which king was said to have had a red-hot poker thrust into his anus?

In 1307, Edward II (1284–1327) succeeded his father, Edward I, to become King of England. Despite being homosexual, Edward married Princess Isabella of France; however, this didn't stop him from openly kissing and cuddling his good friend, a flamboyant young knight called Piers Gaveston. Isabella was unhappy with Edward's behavior but for many years acted as a loyal and supportive wife, and gave birth to four children.

Edward made Piers the Earl of Cornwall and gave him a prominent place in court. This so annoyed the other earls that, in 1312, they seized hold of Gaveston, accused him of treason, and cut his head off. Edward was deeply upset about this but eventually recovered and turned his affections to the Despensers, both the father and the son, who confusingly were both called Hugh. It was at this point that Isabella's patience gave way.

She traveled to Paris with her son, the future Edward III, and met an English noble called Roger Mortimer. They became lovers, and together they planned, and then successfully carried out, an invasion of England. Victory came quickly, as no one was prepared to fight for King Edward and his friends the Despensers, and so within weeks Mortimer and Isabella were in control. They now ruled the kingdom, and Edward was forced to renounce the throne. Edward was thrown into a dark and damp cell, underneath which the corpses of dead prisoners had been left to rot. He was fed rotten food and dirty water. Isabella and Mortimer allegedly had the King murdered in 1327, at Berkeley Castle, Gloucester, by having a red-hot poker thrust into his anus. However, Isabella and Mortimer announced that Edward had died from natural causes, and they even attended his burial at St. Peter's Abbey in Gloucester.

what is the medicinal use of hangman's rope?

Before the development of modern medicine, many illnesses were treated in ways that we now consider superstitious, wrongheaded, or just plain bizarre. For example, in the 1700s, it was believed that the way to cure a headache was to tie a piece of hangman's rope around your skull. Luckily, this was a period in which there were regular public hangings, which meant there was an abundance of hangman's rope, thus making the hangmen a good second income. The following are a number of other strange remedies from the past:

cow dung

Cow dung was a traditional cure for skin conditions, such as ulcers, inflammation, abscesses, and boils.

urine

The book *Old-Fashioned Remedies* recommends, "When your feet are sweaty and aching, or if you have got athlete's foot, blisters or bruises, soak your feet in a bowl of hot urine."

frog spit

It was widely believed that a wart could be cured if it was rubbed with frog spit or snails, which should then be impaled on the thorns of bushes. Licking the eyes of a frog was recommended for eye complaints.

insects

To help treat a fever it was recommended to eat wood lice that were rolled up into balls, known as "pill bugs." Consuming a spider was also believed to help bring down a fever, as was keeping a spider in a bag and hanging it around the neck.

mercury

Mercury was recommended for syphilis sufferers, and was even sometimes injected into the penis of sufferers. Alternatively, it might be mixed with pig fat, among other substances, and then applied to the genitals. However, this "cure" would have been fairly disastrous, as mercury is extremely poisonous, and so it would have caused fever and sores in the mouth and nose, and also made the poor patient's hair and teeth fall out.

plague

There were numerous daft suggestions to help ward off the plague, including the following:

- Catch syphilis
- Keep dirty goats in your home
- Fart into a container, and release it when the plague is near
- Smoke tobacco—children as young as three were ordered to smoke, for their own safety
- Press a dead, dried toad against swollen lymph glands

who did henry VIII order to be boiled alive?

John Fisher (1459–1535) was the Bishop of Rochester during the reign of Henry VIII. After Fisher allegedly treated his cook poorly, the man, whose name was Richard Roose, decided to take revenge, which he did by putting a special herb, or some believe poison, into the bishop's feast one night. The herb was supposed to give the guests diarrhea. Unfortunately, Roose seems to have overdone it, as fifteen guests became ill, and two of the guests suffered such bad diarrhea that they died. Roose was arrested and was found guilty of casting "a certayne venym or poison into the yeaste or barme wyth whych porrage or gruel was mayde for the famyly of the Byssopp of Rochester and others." The cook was tried for murder and was set to be hanged, until the outraged Henry insisted on a more fitting punishment, and ordered that Roose be boiled alive in his own pot. For the next five years, the punishment for poisoners was to be boiled alive.

Roose was boiled to death in 1532. He was forced into an iron caldron, and a fire was lit underneath; it took him two hours to die. A historian wrote, "Because of its novelty, the event attracted a far larger crowd than attended the more commonplace executions of hanging or burning." Later executions by this method were designed to be more humane—they at least placed the criminal in the caldron when the water was boiling hot!

However, a few years later, Henry VIII ordered the execution of sixty-six-year-old Fisher, for refusing to acknowledge his king as the Head of the Church of England. Fisher was put on trial and charged with treason, a crime that carried the death sentence. Until his execution, he was imprisoned in the tower. One day his cook decided not to prepare him a meal, and when Fisher asked why, the servant replied: "It was common talk in the city that you should die, and so I thought it needless to prepare anything for you." Shaking his head, the bishop replied: "Well, for all that, thou seest me still alive; so whatever news thou shalt hear of me, make ready my dinner, and if thou seest me dead when thou comest, eat it thyself!"

In 1535, Fisher was beheaded, and his head was parboiled and stuck on a pole at London Bridge. After a fortnight, it was thrown into the Thames, its place being taken by that of Sir Thomas More.

which gruesome documentary featured a young woman drilling into her own brain?

Boring a hole into the head, which is also known as "trepanning" or "trepanation," is a practice that was widely used for many centuries as a cure for madness and many other disorders. The ancient Greeks used trepanation as the cure for almost all head injuries. Ancient Egyptians performed trepanning operations to alleviate pressure on the brain, which might be caused by conditions such as stroke. In 1965, after years of experimentation, Dr. Bart Hughes bored a hole into his own skull using an electric drill and a scalpel, having first administered himself with a local anesthetic using a hypodermic needle. Afterward, Hughes became a keen advocate of the benefits of his new state of consciousness. People who have been trepanned report an increased ability to concentrate, and generally say they feel good for a long while afterward.

Dr. Hughes became a close friend of a young woman called Amanda Feilding, who decided to film her own trepanation in a gruesome documentary called *Heartbeat in the Brain*. In the film, we see her make a cut in her head with a scalpel and then calmly bore through her skull using an electric drill. There is one particularly grisly moment when blood spurts out of Amanda's head, and she turns smiling toward the camera. Amanda, who twice ran unsuccessfully as a UK parliamentary candidate, wanted the public to receive free trepanation on the NHS (National Health Service).

what was found in genghis khan's hand when he was born?

Genghis Khan (ca. 1162—1227) was a fearsome Mongolian warrior, and his Mongol army killed hundreds of thousands of people across Asia and Europe, and conquered more land than anyone else in history. Genghis was born with a big clot of blood in his right hand, which the Mongols believed to be a sign that he would be a powerful warrior.

Genghis died in his sixties in the year 1227, and, as legend has it, his remains and treasures were taken two hundred miles by his most loyal troops, who killed every living thing in their path, to prevent anyone from ever discovering the grave site. Horses trampled over his grave to further shield it from discovery, and the party responsible for his burial were also killed. Seven hundred years later, the exact location and contents of his tomb are still unknown.

After Genghis's secret burial, his sons and grandsons continued to expand the Mongol Empire. One of Genghis's sons, Toluy, was a cruel and sadistic madman just like his father. After massacring many people in a city, he was told that some people had escaped death by hiding among the bodies and pretending to be dead. Determined that this mistake should not be repeated, on his next massacre he ordered all the corpses' heads to be cut off and arranged in pyramids—one for men, one for women, and one for children. Other Mongol warlords thought this was a good idea, and so began a tradition of warriors building large towers of severed heads.

A study published in 2003 estimates that 8 percent of men across a large portion of Asia are descended from Genghis Khan. This astonishing revelation is based on DNA testing.

who was typhoid mary?

Worldwide, up to 17 million people suffer with typhoid fever annually, with an estimated 600,000 deaths. Typhoid fever is a bacterial infection caused by the bacteria *Salmonella typhi*. It is form of food poisoning, which these days is usually caused by shellfish grown in water contaminated with sewage, or by eating food that has been contaminated by the hands of another sufferer. Typhoid victims suffer symptoms including fever, severe headache, nausea, diarrhea, and abdominal pain, and the disease can be fatal. Fortunately, anyone traveling to a country where typhoid fever is a problem can get vaccinated.

The first identified typhoid carrier in the United States was an Irish immigrant called Mary Mallon (1869–1938). Mallon is believed to have been responsible for hundreds of people contracting typhoid, and was consequently dubbed "Typhoid Mary." She was born in Ireland and emigrated to the United States when she was fourteen years old. By 1906, Mary had worked as a cook for many wealthy families in New York. In 1906, she was working for the Warren family. A month later, half of the members of the household began to get ill, including Mrs. Warren, two of her daughters, and three of the servants. Mary did not become ill, and left the house just three weeks after the typhoid outbreak.

A man called George Soper, a sanitary engineer, was hired to find the cause of the outbreak. After a thorough investigation he came to the conclusion that it was Mary who was responsible for the outbreak. He confronted Mary, who angrily denied ever having typhoid fever. Fortunately, her employment agency gave him a list of some of Mary's previous employers. He subsequently discovered that seven of the eight families for whom she'd worked during the previous ten years had suffered an outbreak of typhoid while Mary worked as their cook. Finally, in 1907, the police forcibly took Mary to a hospital, where she was found to be a carrier of typhoid. She was detained for three years but was released in 1910 after promising she would not work as a cook or with food in general. However, after a while Mary disappeared, and in a hospital in New York there was an outbreak of twenty-five cases of typhoid fever. Three months prior to the outbreak, a new cook called Mrs. Brown had been taken on—this of course turned out to be Mary Mallon. She again disappeared but was later caught and detained in a hospital until her death in 1938. Throughout her life, Mary continued to vehemently deny that she had ever carried or transmitted typhoid, despite all the medical evidence to the contrary.

why did aztec priests rip out the still-beating hearts of their victims?

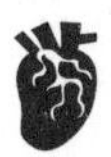

The Aztecs of Mexico, whose empire flourished from 1427 to 1521, are famous for their human sacrifices. The Aztecs were a warlike people, and their armies set out to invade and take over all neighboring areas. When they did so, they would take back thousands of captives to their capital city, Tenochtitlan, where they would be imprisoned in wooden cages and well fed, until their time came to be killed as human sacrifices.

The Aztecs were a polytheistic people, which means that they believed in many gods, rather than just one. They believed that these gods provided them with rain, sunshine, and everything else they needed to live, and so in return they offered human sacrifices. In fact, they believed that the gods would be angry and might destroy the world if they didn't offer sacrifices. About 20,000 people a year were believed to be sacrificed by the Aztec priests. However, in 1487, an unprecedented 80,400 captives were sacrificed in the emperor Ahuitzotl's reconsecration of the Great Pyramid of Tenochtitlan. The Aztec priests painted their bodies black in order to symbolize religion and war. Their hair was matted with blood from human sacrifices, and they filed their teeth to sharp points. They would wrap themselves in cloaks made from the skins of their previous victims.

Most prisoners were sacrificed by having their hearts cut out with a sharp knife, while other captives had their heads chopped off. However, there were also much slower and nastier ways of being killed, which included being thrown into a lake and left to drown, or being skinned alive. If you happened to be unhealthy, scarred, diseased, fat, or ugly, you might consider yourself lucky, as the Aztecs would not choose you for sacrifice. Instead, you would have been forced to work as a slave.

On the big day, the prisoners would be given a herbal potion to make them feel drowsy. Soon afterward, they were walked up the steps of the temple at the center of Tenochtitlan, where the victim would be laid onto a stone slab. The prisoner's abdomen would be sliced open with a knife made of flint, and then the priest would tear out the still-beating heart from the victim's chest before offering it up to the gods. Next they would cut off the head, which would be displayed in a skull rack, along with hundreds of others. The prisoner's blood was poured down the temple steps, and the limbs were given to the lucky warrior who had captured the human sacrifice. Naturally, he would make a delicious meal out of the remains; in particular, the hands and thighs were regarded as delicacies.

Between 1519 and 1521, Tenochtitlan was invaded by groups of Spanish soldiers, who used guns, horses, and vicious dogs to fight the Aztecs. However, their most effective weapon turned out to be a tiny germ. One of the soldiers was carrying a highly infectious viral disease called smallpox, and in the course of the invasion, he inadvertently managed to spread the deadly smallpox germs to the Aztec people. Having never been exposed to smallpox before, they had no immunity, and thousands of Aztecs died as a result. In fact, it's believed to have killed approximately half the Aztec population in less than six months. In 1521, the Spaniards looted Tenochtitlan of its innumerable treasures, and its buildings and temples were burned or pulled down. A new capital, Mexico City, was built on its site.

which infection causes bleeding from the eyes?

One of the most frightening viruses to come out of the African subcontinent is the Ebola virus. This disease is poorly understood, and the causes of an outbreak are unknown. However, bats have been implicated, as it seems possible that they may be able to carry the disease without displaying any symptoms. Outbreaks of Ebola first occurred in the Sudan and Congo (then Zaire) in 1976, and since this time more than one thousand two hundred people infected by the virus have died from it.

The Ebola virus causes a number of horrible symptoms, including high fever, headache, severe pain, red eyes, vomiting, and diarrhea. As the disease progresses, it can cause bleeding from every opening in the body. This is because the virus causes the blood vessels to break down, allowing blood to seep out into the body. Blood will then flow freely from the nose, eyes, ears, gums, vagina, and rectum. Victims die of organ failure or excessive blood loss, and most of those infected die within two weeks.

Ebola is a highly contagious disease, and is generally transmitted via person-to-person contact through infected body fluid, or through using contaminated needles, but in certain stages it can be transmitted by as little as a handshake. Close contact with the corpses of victims can also spread the virus if they are handled without the wearing of suitable protective clothing. Fortunately, Ebola seems unlikely to develop into a pandemic, because it kills its victims faster than it can spread, and so it burns itself out before traveling too far.

why might the mention of a goat's tongue have struck fear into the hearts of medieval prisoners?

In the Middle Ages, there were a range of clever and inventive torture methods designed to punish or force a confession out of you. In a punishment known as the "goat's tongue," or "tickling torture," the victim's feet would be placed into wooden stocks and covered with salt or a sweet substance.

Goats love salt, and for hours will happily continue to lick feet that are covered in it. Although the sensation starts out as initially being little more than a tickle, pain soon follows, as the goat's rough tongue causes the soles of the feet to blister and become sore. By the time the goat's tongue reaches bone, a confession is usually forthcoming. Vlad the Impaler of Transylvania enjoyed using this torture on Turkish prisoners. Salt water would be continually dripped onto their feet from above, and enthusiastic goats would be released.

Another unpleasant torture technique of the Middle Ages was "the rack." This nifty device was used to extract confessions, and was introduced to the Tower of London in about 1420 by the Duke of Exeter. The torture rack consisted of a strong wooden frame fitted with a large roller at each end. The prisoner would be laid on the frame, with ropes attached from the rollers onto his ankles and wrists. If the prisoner did not confess to his crime, the rollers would be turned, and the body slowly stretched, pulling the limbs in opposite directions until the joints of the limbs would eventually become dislocated, resulting in excruciating pain. This method was also used with great relish during the Spanish Inquisition.

Guy Fawkes (1570–1606) was put on the rack in 1605, and was tortured for ten days at the Tower of London until he confessed the names of the other members of the Gunpowder Plot. He was so severely injured that he had to be carried to the scaffold for his execution.

During the reign of Elizabeth I, many Catholic priests were executed as traitors. English Jesuit Edmund Campion (1540–81) was jailed for heresy, after refusing to recant his Catholicism. He was tortured but, even when iron spikes were driven under his fingernails and toenails, he refused to confess. He was later placed on the rack, which caused extreme agony, but he still refused to confess. It is said that when he came off the rack, he was 4 inches (10 cm) taller than before. Campion was given a final chance to renounce his Catholic faith in exchange for his freedom, but his loyalty to the Pope proved too strong. In 1581, Campion was executed as a traitor and hanged, drawn, and quartered at Tyburn, London.

Thumbscrews were handy, pocket-sized contraptions that were highly effective at inflicting excruciating pain. The victim's fingers were placed inside the instrument and slowly crushed, as the torturer turned the handle on top. Some cunning thumbscrews even had little studs inserted inside them, to increase the suffering of the victim. An overenthusiastic torturer would sometimes crush the finger to a pulpy mess! In Scotland, thumbscrews were known by the rather cute name of "thumbkins," and were used until the end of the seventeenth century.

The same instrument was also used to crush victims' toes, and even bigger variants were used to crush victims' feet, knees, and elbows. Another devious device was the head crusher, which, as you might imagine, was designed for applying pressure to the head.

what does it mean to be hanged, drawn, and quartered?

From the thirteenth century until the 1800s, the penalty ordained in England for high treason was the extremely gruesome method of being hanged, drawn, and quartered. The traitor would be dragged to the scaffold on a wooden frame (called a hurdle) by a horse—this is one possible interpretation of "drawn." Then he was hung by the neck until near to death. Next, he was taken down from the gallows, cut open, disemboweled (his internal organs removed), and his genitals were cut off, which all occurred while he was still alive. He would watch as his body parts were tossed and burned in a fire. The body would then be cut into four parts, that is to say "quartered," before finally being beheaded. Typically, the various parts would then be put onto spikes and displayed publicly.

In 1757, Robert-François Damiens, a former soldier in the French army, was executed for the attempted assassination of Louis XV. The execution was public, with a large crowd attending the particularly gruesome scene. Prior to the execution, Damiens was kept strapped to an iron bed, to prevent him from committing suicide, although he did manage to bite off his tongue, in the hope that he might bleed to death. To stop him from doing this again, the guards pulled out his teeth. Then Damiens was tied to a scaffold and burned with red-hot pincers on his chest, arms, thighs, and calves. As he was being burned, Damiens "gave a very loud and continuing cry, which was heard at a great distance from the place of execution." The hand in which he had held the knife, with which he had attempted to kill the king, was then burned off, using sulfur and burning oil. Next, molten wax, lead, and boiling oil were poured into his wounds. Ropes were then tied to his arms and legs, and attached to horses, which were made to run in opposite directions, so that his limbs would rip off. Unfortunately, his joints wouldn't tear, so the executioner had to cut through them with a knife. Damiens was seen to look at his severed limbs before dying. As soon as he appeared to be dead, his remains were thrown onto a fire.

Damiens's friend, the infamous lover Casanova, who witnessed the execution, wrote in his memoirs, "I was several times obliged to turn away my face and to stop my ears as I heard his piercing shrieks, half his body having been torn from him."

Another brutal and common method of punishment was known as "the wheel." The wheel always killed its victim, but it did so very slowly and painfully. One variation of this method involved tying a victim to a large wooden wheel, like a cart wheel, so the limbs were hanging over its sides. The executioner would then use a metal bar or pole to break the limbs, until they became so pulverized that they would fit nicely around the outside of the wheel. This allowed the executioner to easily roll the criminal around town. Sometimes, through the openings between the spokes, the torturer would break each bone separately, finally killing the prisoner by hitting him across the chest. However, occasionally the executioner would break the man's ribs, and the wheel would then be mounted on a tall pole, so that birds could feed from the still-living prisoner. He would remain there until his death.

The English were particularly fond of another brutal punishment, the pleasant sounding "*peine forte et dure*," or "pressing to death." English common law stated that a prisoner had to plead either guilty or not guilty, and until a prisoner gave a plea, a trial couldn't be held, so it was common for those arrested to be "pressed to plea." It meant literally pressing the accused's body with heavy weights until he gave a plea. If he refused to plead, the prisoner was laid on the ground, naked, and his arms and legs were drawn with cords fastened to several parts of the room. Then increasingly heavy weights of iron, stone, or lead would be piled on his chest.

One reason prisoners were reluctant to plead was that if they admitted guilt, they would be hanged, which in those days was a lingering and painful death, although pressing wasn't much better. More important, as a convicted criminal, all their possessions would be taken away, whereas if they said nothing, even though they would be "pressed" to death, they would die without being convicted, thereby saving their families from a life of poverty.

In the late 1600s, an eighty-year-old farmer called Giles Corey was accused of witchcraft during the Salem witch trials. He refused to plead, and so chose to be pressed to death, as he wanted to ensure that his wealth was passed on to his family. Now and then the torture was halted to give Corey the chance to plead, but all he said was "More weight!" He was tortured for two days, before finally suffocating to death.

why was mary tudor nicknamed "bloody mary"?

When Mary Tudor (1516–58) became queen in 1553 she was determined to return England to Catholicism. In 1554, Protestants were horrified when she married her cousin, the Catholic King Philip II of Spain, and ordered hundreds of Protestants to either convert to Catholicism or be burned to death. She ordered 288 Protestants to be burned at the stake, earning her the nickname "Bloody Mary." She believed that fire burned away all evil, allowing the purified souls to go straight to heaven. To make death quicker, and a little kinder, the executioners would sometimes strap gunpowder to the victim's body.

When Mary came to the throne, two Protestant bishops called Hugh Latimer and Nicholas Ridley were arrested and tried for heresy. At the trial, the accuser pointed to Latimer and Ridley and demanded, "Swear allegiance to the pope, confess your heresies, and you will live." Both men refused, and in 1555, they were burned together in the center of Oxford, in front of a large crowd. The following famous excerpt from John Foxe's *Book of Martyrs* (1583) describes how they were burned.

> So they came to the stake. Dr. Ridley, entering the place first, looked toward the Heaven. Then, seeing Mr. Latimer, with a cheerful look, he ran and embraced him saying, "Be of good heart, brother, for God will either ease the fury of the flame, or else strengthen us to endure it.
>
> Dr. Ridley's brother brought him a bag of gunpowder and tied it about his neck. Then they brought a lighted faggot and laid it at Dr. Ridley's

> feet. Mr. Latimer said, "Be of good comfort, Dr. Ridley, we shall this day light such a candle, by God's grace, in England, as I trust never shall be put out." He then cried out, "Father of Heaven, receive my soul."

Other unfortunate victims included Lady Jane Grey (1537–54) and her husband, Guildford Dudley, who were tried for high treason in 1553. Lady Jane had to watch her husband's execution. He was crying as his head was chopped off, and Lady Jane's execution followed shortly after. She was buried underneath the floor in the Tower of London, and her ghost is said to walk the Tower to this day. According to the legend, her ghost appears every year on the anniversary of her execution as a white shimmering figure that floats, strolls, or glides around the Tower, and then slowly withers away.

which sultan hacked off the heads of men with fat necks?

Murad IV (1612–40), also known as "Murad the Cruel," became ruler of the Ottoman Empire when he was only five, an enormous area that stretched from Jerusalem to Constantinople, in 1623. During his younger years, his dominating mother, Sultana Kosem, effectively ruled through him.

When he grew older, Murad was mean and sadistic.

He would sometimes walk through streets accompanied by an executioner, and if he didn't like the look of a person, or if they acted in what he considered to be a strange way, he would have them executed. At one point, anyone considered to have a fat neck was killed on the spot. Another time, while out hunting, he came across some women happily singing while having a picnic. The noise displeased him and so he ordered them to be drowned. He also enjoyed shooting servants and members of his harem.

Murad hated smoking, and believed it made people infertile. In fact, he was so opposed to it that he issued a law stating that anyone found smoking would be killed, either by beheading or being hanged, drawn, and quartered. Astonishingly, thousands continued to smoke, and many met a grisly demise. Murad's cruel example was followed in Persia, where merchants caught selling tobacco were executed by having molten lead poured down their throats.

In 1640, at the age of twenty-eight, he died of cirrhosis of the liver, a result of drinking too much alcohol. Ironically, during his time in power, he had also banned alcohol as well as tobacco. On his deathbed he commanded that his brother Ibrahim should be executed, and wanted his head cut off and brought to him as proof of his death; however, Murad died soon after and Ibrahim became the new sultan, but unfortunately he was just as mad as his brother Murad.

how long does a head remain conscious after being guillotined?

The guillotine is named after a French physician, Joseph-Ignace Guillotin, who, ironically, didn't agree with the death penalty. In 1789, he suggested using this machine to execute criminals, as it was a less painful method of execution than other methods being used, such as hanging. He was horrified that his name was given to it, and when he died in 1814, after the government refused to change the name of the device, his children changed their name so as not to be connected with it.

In France, during the time of the guillotine, doctors carried out studies to find out if a person's head could remain alive after being decapitated. To research this, the doctors would ask those about to be executed if they would try to keep blinking their eyes after their head had been chopped off. According to reports, some of the heads continued to blink for up to thirty seconds after being cut off, although it is unclear whether this was due to the person blinking on purpose or simply an involuntary reaction.

In late 1905, a French physician named Beaurieux carried out some experiments on the head of a prisoner named Languille immediately after his head was chopped off. He described his experiments: "Here then is what I was able to note immediately after the decapitation: the eyelids and lips of the guillotined man worked in irregularly rhythmic contractions for about five or six seconds." He went on:

> . . . I called in a strong, sharp voice, "Languille!" I then saw the eyelids slowly lift up. . . . Next Languille's eyes very definitely fixed themselves on mine and the pupils focused themselves. . . . I was dealing with undeniably living eyes which were looking at me. After several seconds, the eyelids closed again. . . . It was at that point that I called out again, and, once more, without any spasm, slowly, the eyelids lifted and undeniably living eyes fixed themselves on mine with perhaps even more penetration than the first time. . . . I attempted a third call; there was no further movement—and the eyes took on the glazed look which they have in the dead. . . .

The guillotine was last used in France to execute a Tunisian immigrant for the rape and murder of a young girl in 1977. It was banned in 1981 by President François Mitterrand, with the outlawing of capital punishment.

what is the worst sound in the world?

It's official, the most unpleasant sound in the world is that of a person vomiting. A survey was carried out to identify the world's worst sound, and, perhaps unsurprisingly, vomiting was voted as the sound that made most people want to cover their ears. The sound was re-created by an actor using a diluted bucket of baked beans to mimic the sound of vomit cascading into a toilet.

The study, set up by Trevor Cox, a professor of acoustic engineering at Salford University, asked for people's opinions on thirty-four sounds, which were hosted on a website, in the hope of learning what makes certain noises so universally disgusting. The survey attracted more than 1.1 million votes from around the world. People listened to sounds such as a dentist's drill, fingernails scraping down a blackboard, and an aircraft flying past, before rating them in order of offensiveness. Microphone feedback was judged to be the second most unpleasant sound, while joint third were the sounds of babies crying and the scrapes and squeaks of a train on a track. Surprisingly, the sound of snoring was deemed only the twenty-sixth most objectionable.

what is gas gangrene?

One infection you definitely do not want to suffer from is gas gangrene, which is an even more severe form of gangrene. Although rare, it carries a high risk of death and is extremely painful. It typically results from deep wounds becoming infected, usually through injury or surgery. After an injury to the body, there may be a poor supply of blood to the tissues, and as blood carries oxygen, this lack of oxygen can cause the flesh to die.

Gas gangrene occurs as a result of infection by clostridium bacteria. Under low-oxygen conditions, these bacteria produce toxins that cause tissue death and other associated symptoms. After an injury, clostridium spores can find their way into the deep wound, germinate, and grow in the dead tissue, causing symptoms such as swelling and severe pain in the affected area. One very unpleasant symptom is that the bacteria produce gases under the skin. This can cause a crackling sensation when the skin is touched, known as "crepitation," which is due to the movement of these gas bubbles.

The skin over the wound may rupture to reveal necrotic (dead tissue) muscle, which will have turned black, and will emit a foul-smelling watery or frothy discharge. If sufferers of gas gangrene are treated quickly, 80 percent of patients will survive.

However, if it affects areas such as the abdominal wall or the bowel, the prognosis is poor. The most common treatment involves surgically cutting out the affected area, or in many cases amputation of the limb.

is it possible to catch gonorrhea from an inflatable doll?

Gonorrhea is a sexually transmitted disease caused by a bacterium, *Neisseria gonorrhoeae*, which can grow and multiply easily in the warm, moist areas of a person's sexual organs. This bacterium can also grow in the mouth, throat, eyes, or anus. Symptoms of gonorrhea in men include a burning sensation when urinating, and a white, yellow, or green discharge from the penis. Sometimes men with gonorrhea also get painful or swollen testicles.

In 1993, the following story was published in a medical journal called *Genitourinary Medicine*. A Norwegian physician, Harold Moi, took a job working for a venereal diseases clinic in Nuuk, Greenland. One of his patients was the captain of a trawler who arrived at the clinic with symptoms similar to those of gonorrhea. A blood test confirmed that he was indeed suffering from gonorrhea; however, the origins of the disease were something of a mystery.

The skipper had been at sea for three months, and the symptoms had appeared toward the end of his voyage. As gonorrhea takes no more than thirty days to appear, it was evident that the man must have caught the disease while at sea. However, the boat's crew were all male, and the skipper insisted that he had never had a homosexual experience, so how could he have contracted it? The mystery was resolved when the captain confessed that one night he had gone to a crewman's cabin and borrowed the man's inflatable doll without asking his permission. A few days later, he had noticed the first symptoms of the disease.

Reportedly, the owner of the doll had used it shortly before the skipper, and he was also found to have gonorrhea; indeed, he had it much worse than the skipper. It seems that he had caught the disease from a girl he'd slept with before going to sea and then inadvertently passed it on to his captain via the inflatable doll.

who drinks fresh cow's blood?

Like farmers the world over, the Maasai people of northern Tanzania keep cattle for their milk, their meat, their hides, and to help with the work of the farm. However, the Maasai also have another, more unusual use for their cattle. To give them strength, they also drink cow's blood. However, to do this, they don't need to slaughter the cow. Rather, they simply drain the blood by cutting one of the cow's veins and allowing the blood to pump out into a cup, resulting in a fresh supply of warm blood. Sometimes the Maasai drink the blood as it comes; at other times, they mix it with sugar, or simply let it clot, and then divide it into portions. They also give a mixture of cow's milk and blood to the sick, to help them recuperate, and to new mothers, so that they might regain their strength after giving birth. Apparently, the cows are usually fairly untroubled by being tapped in this way, and are unharmed by the process. A similar practice was carried out by the Mongol warriors of Genghis Khan, who sometimes drank the blood of horses while engaged in battle.

If the idea of eating clotted blood strikes you as bizarre, you might consider the traditional English dish of black pudding, which is a staple ingredient of a cooked breakfast in many parts of the country. Black pudding, as made in the UK, is a blend of onions, pork fat, oatmeal, flavorings, and clotted blood, usually pigs' blood.

chapter six

the final chapter

who gives massages to dead people?

What could be nicer than a long, soothing massage? Massages help to aid relaxation and tease out our bodily aches and pains. However, massages are not restricted to the living. Dead people also receive massages, but for somewhat different reasons. Massage of the dead serves as part of the modern embalming process. Embalming has been carried out for a wide range of reasons down through the ages, but these days it is mainly designed to preserve the body so that it can be viewed at a funeral ceremony.

The ancient Egyptians were so skilled at embalming bodies that thousands of them have been found intact over the last millennium. They believed that a person's spirit would return to the body after death, which meant the body had to be kept in good condition and not allowed to deteriorate. The ancient Egyptians used spices, herbs, and salt to preserve dead bodies. First, they would remove the body's internal organs through the anus and scoop the brain out through the nostrils using a metal hook. Although methods varied, many Egyptian embalmers would then use linen strips soaked with resin to pack into the empty skull. The organs would be cleaned and then mixed with spices, herbs, and resins before being put back into the body or placed into jars. The inside of the body was washed with spiced wine and allowed to dry out by being immersed in salt. After a few months, the body would be cleansed, stuffed, and anointed with oils and resins, to deter insects. Finally, the incisions were closed up, and the body was bandaged; the bandages were held together with gum or glue, resulting in a beautifully preserved mummy.

In modern times, the embalming process is generally referred to as "cosmetic embalming," and its main purpose is to improve the appearance of the body and prevent deterioration leading up to the funeral. Modern embalming fluid consists of a solution of formaldehyde, often containing pink dye, that is injected into the blood vessels, usually via the carotid artery, which is found in the neck. First, the embalmer will insert a tube into a vein so that blood can be drained from the body. Then the embalmer massages the body's limbs and hands to help the blood flow out, clear any blood clots, and encourage the flow of embalming fluid as it moves through the body.

The pink dye added to the embalming fluid helps to add color to the skin, to give a more healthy, life-like appearance. The fluid also helps to plump out the deceased's features, so that they look less tired and drawn. The embalmer will also arrange the dead person's hair and use makeup to make them look their best for when they receive visitations. Massage cream may also be applied to the face and hands to prevent dehydration, and also to aid the massage of the tissues during the embalming process.

what happens to the body after death?

If a corpse is exposed to the elements in hot conditions, a body can become a skeleton in about a month; however, if it is buried 6 feet (2 m) underground, even without a coffin, it can take more than ten years. Of course, if dead bodies are not embalmed, they quickly begin to putrefy.

Upon death, the muscles relax and blood pools in the body parts closest to the ground, giving the underside skin a darkened appearance. The top of the body will become a grayish-white color. Within minutes, flies will pitch on the body and can smell the corpse's odor from at least a mile away. They lay thousands of eggs, which soon develop into maggots. The maggots will enthusiastically feed on the corpse's organs and tissues, and can reach adulthood in three weeks. Insects and larvae that feed and inhabit a corpse are useful to murder investigators, as they can help them to determine the date of a person's death.

Stiffening of the corpse occurs within six hours

of death, in a process known as "rigor mortis," and after a day the body slowly relaxes again. From forty-eight hours after death, bacteria living in the body, especially the colon, begin to affect the decomposition of the body. Beginning on the lower abdomen, the skin becomes a greenish color, which spreads to the rest of the body, which by now will be producing a horrific stench. Gas produced by the bacteria causes the body to bloat and turns the skin from green to purple and then to black. The buildup of gas forces the intestines to be pushed out of the body, through the vagina or rectum. This gas also produces a foul stench, as it forces copious amounts of putrid bloodstained fluid to come out of the mouth, nose, and other bodily orifices. That is why coffins are made with lids that can "burp" so they can withstand the pressure of these gases.

Within a week, the skin may develop putrid blisters that burst. Soon, large patches of skin are shed in a process that is known as "skinslip"—the skin of the hand can slough off like a glove. Some of the organs, such as the eyeballs and the brain, liquefy. After a few weeks, teeth, nails, and hair loosen and may fall off, and much of the body turns into a semiliquid state. Later, the abdomen and chest areas burst open and the body dissolves.

Historically, many cultures have been petrified of the dead returning to haunt them, and these fears have formed the basis of many of today's burial rituals. The Saxons used to cut off the feet of their dead, to prevent them from walking around at night. In northern Europe, people went to extreme lengths to make sure the dead remained dead. Chopping off the head and feet of the corpse and tying its hands together were commonplace. They took no chances on the risk of being haunted; digging a 6-foot-deep hole would help to ensure this. Lids of coffins were nailed on tightly, not to prevent the corpse from falling out but to keep it from jumping out at one of the mourners. After the coffin was lowered into the grave, a large rock was placed on top, followed by dirt, and then another big slab of stone was dragged on top, to ensure that the dead body could not escape. This second slab of rock eventually evolved into the tombstone.

In Britain, the custom of wearing black and wearing a veil over one's head wasn't originally intended as a sign of respect for the dead but rather was meant to help hide the identities of mourners from any bad spirits hanging around in the graveyard who might otherwise cause them harm.

why did undertakers prick holes into corpses?

After a person dies, the bacteria that are already present in the body start to dramatically increase in number, as the immune system can no longer keep them in check. As these bacteria munch away at the corpse, they produce a huge buildup of gas inside the body. The buildup of these gases causes the body to change shape in extraordinary ways: the eyes and tongue protrude, the abdomen distends, the lips swell, and the breasts and genitals balloon in size.

However, this buildup of gases is not just unsightly—it can also be dangerous. If the body was to burst and discharge its gases, the noxious fumes would cause people to vomit or even faint. Consequently, undertakers and gravediggers developed various techniques to deal with these risks. Gravediggers, when they opened a coffin, would immediately flee to a safe distance and stay there until the corpse's gas had been sufficiently dispersed. Another custom was to burn paper or straw in graves, while some men believed a mouthful of garlic would protect them from the putrefaction of the corpses. Cases have even occurred of gravediggers being suffocated while thrusting a pickax into an abdomen distended with gas.

During the 1800s, coffins used to be "tapped" to allow the escape of gas. This tapping process involved boring a hole into the coffin, that permitted the gas to pass through it. The gas would then be ignited, producing a flame that lasted from around ten minutes to half an hour. On occasion, the buildup of corpse gas could be so intense that it caused the coffins to bulge, or even explode.

Another method of releasing gas from bloated corpses involved undertakers pricking tiny holes into the body and then holding a candle to the openings. Long blue flames would appear, which were fueled by the gases escaping from the corpse, and these flames would continue to burn until all the gas was gone.

what was a "waiting mortuary"?

During the seventeenth and eighteenth centuries, the rudimentary nature of medicine meant that there was a risk of people who were comatose or otherwise incapacitated being misdiagnosed as dead. Consequently, many people in Europe and America were terrified of being prematurely buried. This led to a plethora of pamphlets and books on the phenomena of "apparent death" and premature burial.

A German physician named Christoph Wilhelm Hufeland also took a great interest in premature burial, and believed that putrefaction was the only certain sign of death. He also thought that when a person died they first entered into a state of deep unconsciousness, or "death trance," which was indistinguishable from real death, and which could last for a long while. However, Hufeland argued, the death trance was not always fatal. A person could sometimes awaken from this state, which meant that there was a risk that they might be buried alive.

To solve this problem, in 1791 Hufeland built a "waiting mortuary" in Weimar, Germany, which consisted of a "corpse chamber" with eight beds for the deceased. A fire was kept continually alight, and steam from boiling water was fed through underground pipes to the corpse chamber; this was designed to help ensure that any bodies that actually were dead would rot quickly, and thus demonstrate the fact of their death. To help disguise the smell of rotting flesh, bouquets of flowers were placed around the beds to perfume the air.

Additionally, there was an attendant employed to look out for signs of life, in case any of the bodies turned out to have merely been in a death trance. Another gruesome and smelly job of the attendant was to clean up the mess associated with exploding corpses. Each of the bodies had a bell attached to its fingers, which the attendant had to listen out for. According to reports, these bells would often ring, because the corpses would move and expand as they decomposed, but no one sent to a waiting mortuary ever woke up.

However, the waiting mortuaries turned out to have another, incidental benefit. As well as removing the risk of premature burial (even though they appear to have had no meaningful success in this area), the waiting mortuaries also removed the corpse from the home of the grieving family. This idea took off, and waiting mortuaries became popular in Germany in the 1800s, providing a place where the corpses would be kept until they showed signs of putrefaction. These establishments were staffed by a matron, nurses, and porters. A doctor was also kept "on call," should any of the patients show signs of life.

In Frankfurt am Main, a wealthy man, terrified of being prematurely buried, donated money to build a luxury waiting mortuary. It consisted of two wards, with sixty-six beds in total, which were constantly observed by nurses and porters. A doctor also made regular daily visits. All the dead bodies wore white gloves, and their hands were attached to a powerful gong by a piece of string, so if there was any movement, the gong would make a noise.

After the discovery of the heartbeat as a clear sign of life or death, the European fear of being prematurely buried diminished from 1846 onward. The last German waiting mortuaries were closed in the 1890s. Further, the development of embalming and cremation in the late 1800s helped to give people peace of mind about being buried alive.

what is a "safety coffin"?

Waiting mortuaries were not the only innovations designed to protect people against the risk of premature burial. From the mid-1800s through the early 1900s, a number of coffin alarm devices were patented. In the 1850s, George Bateson made a lot of money from his very successful "safety coffin," which was marketed as the Bateson's Life Revival Device. The device consisted of a bell that was attached by rope onto the corpse's hands, through the lid of the coffin. If the corpse awakened, it could simply ring the bell for assistance. Queen Victoria awarded Bateson the Order of the British Empire for his services to the dead. During the Victorian age, his invention and other coffin bell alarms could be found in many of the new graves in cemeteries far and wide. However, despite the success of his invention, Bateson was driven insane by his obsession with being prematurely buried. Even his own invention didn't allay his fears, so he rewrote his will asking to be cremated.

how long could a person survive if buried inside a coffin?

It is perhaps reassuring to know that even if a person was to become trapped inside a coffin, they would not have to endure the experience for very long. Trapped inside a coffin, 6 feet (2 m) underground, a person would quickly exhaust the limited air available. In 1862, the German physiologist Johann Ernst Hebenstreit calculated that there would not be enough air in an average airtight coffin to sustain a person for more than one hour. This claim inspired another scientist, a man named Dr. Von Rosser, who, in the late 1800s, carried out an experiment in which he placed a dog in an airtight coffin with a glass lid—the unfortunate animal died within three hours. As a human has roughly three times the lung capacity of a dog, Von Rosser was inclined to agree with Hebenstreit's calculation.

who was the woman who died three times?

Even with all the advances made by medical science today, doctors can still misdiagnose death. In 2000, a documentary called *Premature Burial* was made about "the woman who died three times." By the age of sixty-five, Allison Burchell of Horsham, Sussex, had been pronounced dead on three occasions in her life.

In 1952, when she was seventeen years old, she collapsed unconscious in a cinema and was rushed to the hospital, where she was pronounced dead. However, despite her body being paralyzed, she was able to see and hear clearly, and was aware of all the conversations being held by the nurses as they prepared her body and eventually took her to the mortuary. Thankfully, after about thirty minutes, she recovered and was then subjected to numerous tests, which revealed that she suffered with severe narcolepsy. This disorder has symptoms that include overwhelming drowsiness, sudden attacks of sleep, and cataplexy, which is a sudden loss of muscle control and can cause short-term paralysis. Narcolepsy can leave patients totally paralyzed, even though they are fully conscious and able to understand everything that's going on around them.

A few years later, Burchell suffered another attack and woke up in a mortuary, surrounded by dead bodies. In the 1970s, she had moved to Melbourne and suffered a third attack. Again, she was pronounced dead, but her teenage son convinced hospital staff not to place her body in the airtight refrigeration unit. Burchell described the sensation: "You can see and hear everything going on around you, but there is no way to convey to anyone that you are not dead. It's the most terrifying thing imaginable." Thankfully, Burchell wasn't put into the refrigeration unit, and once again made a full recovery.

why would doctors in the 1700s and 1800s pour warm urine into the mouths of corpses?

A wide range of extraordinary methods was developed to try and ensure that those who appeared to be dead were, in fact, dead before being buried in the ground.

In 1752, Paris physician Antoine Louis suggested blowing tobacco smoke into the corpse's anus. Thankfully, he wasn't proposing to actually blow the smoke himself. Instead, he developed a piece of equipment that resembled a set of bellows with pipes attached, which he enthusiastically demonstrated on corpses in the morgues of Paris. In the 1800s, tobacco enemas were frequently carried out on corpses by doctors in Holland, London, and Germany, but there was never a successful resuscitation reported.

Other methods seemed to simply assume that pain would sort the actual dead from the merely somnolent. An Englishman named Barnett recommended that the arm of the corpse be scalded with boiling water; if a blister resulted, it meant the body was alive. A French clergyman came up with a rather more direct method of ensuring that an individual was dead—he recommended thrusting a red-hot poker into the corpse's anus. In 1854, another Frenchman, by the name of Jules-Antoine Josat, invented an innovative device called the "pince-mamelon," which means "nipple-pincher." As you might imagine, this device's job was to squeeze the nipple very hard, causing any living body to wake up due to the intense pain.

A Danish anatomist named Jacob Winslow (1669–1760) invented numerous methods to ensure that a person was actually dead.

> The individual's nostrils are to be irritated by introducing sternutaries, errhines [things that cause sneezing and produce mucus], juices of onions, garlic and horse-radish . . . The gums are to be rubbed with garlic, and the skin stimulated by the liberal application of whips nettles. The intestines can be irritated by the most acrid enemas, the limbs agitated through violent pulling, and the ears shocked by hideous Shrieks and excessive Noises. Vinegar and salt should be poured in the corpse's mouth and where they cannot be had, it is customary to pour warm Urine into it, which has been observed to produce happy Effects.

Other methods used included poking long needles under the corpse's toenails, pouring boiling wax over its head, slicing its feet with razors, shouting into the corpse's ears, and thrusting a sharp pencil up its nose, and a Swedish writer even recommended that "a crawling insect" be conveyed into the corpse's ear.

In 1892, French physician Jean Baptiste Vincent Laborde advised tongue pulling, about fifteen to twenty times per minute, which was to be carried out for at least three hours following the supposed death. He even invented a hand-cranked tongue-pulling machine for use in the mortuary. He believed this method was the most effective and successful way of resuscitation, as it "stimulated the respiratory reflex" and would help encourage breathing. He noted its successful use in sixty-three cases.

However, some doctors realized that the most reliable way to identify the dead was simply to allow the body to decompose for a few days. In 1740, Winslow wrote, "The onset of putrification was the only reliable indicator that the subject had died." This meant that corpses would be left hanging around in doctors' surgeries or in the home. The putrid smell and abundance of insects hovering around the body was a sure sign that the person was most probably dead. In 1846, a more scientific way of determining if someone was dead was suggested by a French doctor named Bouchut. Stethoscopes had recently been invented, and Bouchut believed that they could be used to determine whether or not the body's heart was beating. If the heart was not beating, this meant the person was indeed dead.

what is the story of sleeping beauty really about?

Children everywhere love the famous tale of "Sleeping Beauty" in which the handsome prince falls in love with the beautiful young girl. However, while many of us think of it as simply a sweet, romantic fairy tale, others derive a rather different meaning. Some psychoanalysts argue that the story of "Sleeping Beauty" is actually a coded reference to necrophilia, the practice of engaging in sexual acts with dead bodies. The key elements of the story, they argue, are that a beautiful, dead princess is aroused from her permanent slumber by the sexual advances of a handsome prince.

In fact, sexual molestation of the dead has occurred throughout history. In some periods, it appears to have been widespread enough for some societies to take active precautions to prevent it. In ancient Egypt, for example, to ensure that corpses were not dug up and sexually abused, the practice was for bodies to be left to decompose for days before being buried, so that they were sufficiently rotten at the point of burial.

At his court-martial in 1849, Sergeant François Bertrand of the French army admitted to numerous sexual attacks on corpses. His sick antics were initially confined to digging up corpses until his bare hands bled, and then masturbating while touching the corpse's intestines with his hand. Eventually, however, he went a step further and had sexual intercourse with the body of a sixteen-year-old girl he had unearthed. He recalls:

> I cannot describe what I felt during that time. But all my enjoyment with living women is nothing as compared with it. I kissed the girl all over her

> body, I pressed her to my heart as though I wanted to crush her; in short, I did everything to her that a passionate lover does to his mistress. Having enjoyed the body for about a quarter of an hour, I cut it up and, as in the case of my other victims, tore out her intestines.

To satisfy his overwhelming sexual appetite, not only was he carrying out his nightly escapades but he was also involved in sexual relationships with a large number of actual living women, several of whom wanted to marry him.

One night, while climbing over a wall into a cemetery, he was severely wounded by a gun booby trap, which had been set up specifically to catch him. Nonetheless, this didn't deter the tenacious Bertrand, and even though he'd been shot, he managed to dig up a female corpse, cut out her genitals, and then slash her thighs. Several days later, a gravedigger overheard a soldier mention that one of his comrades had been accidentally wounded a few days before, and was in the military hospital. This information led to Bertrand's eventual capture and confession, and he was sent to prison.

In the 1930s, a sixty-year-old German man, Count Karl Tanzler von Cosel (who in fact wasn't a count at all, except in his own mind) lived in Key West, Florida. He worked as an X-ray technician in a local hospital, and one day he was given the job of x-raying an advanced case of tuberculosis in a twenty-year-old Cuban woman named Maria Elena Hoyos. He instantly fell in love and became devastated when she died shortly afterward. Her family buried Elena in Key West's City Cemetery, but many years later it was discovered that her body was missing, and investigators were shocked to discover Elena's embalmed corpse in von Cosel's shack, reclining on a large bed. Von Cosel had replaced her rotten eyes with glass eyes, and reconstructed her face, breasts, arms, legs, and trunk—he had even invented a special vaginal tube that allowed sexual intercourse.

Von Cosel confessed to the abduction of the body and spent time in jail. However, many years later, at the time of his death, morticians found an eerie sight propped in the corner of his small shack: a life-sized replica of Maria Elena Hoyos.

what was the gruesome fate of oliver cromwell's head?

Oliver Cromwell (1599–1658) was the leader of the Parliamentarian army during the English Civil War. His army, the New Model Army, defeated the Royalists, resulting the execution of King Charles I in 1649. In 1653, Cromwell was offered the title of king by a council of officers, but he turned it down and instead assumed the title of "Lord Protector" of England, Scotland, and Ireland.

Cromwell died of natural causes in 1658 and was embalmed and buried with great ceremony in Westminster Abbey, alongside past kings and queens. However, Cromwell was evidently extremely unpopular at his death, as Royalist John Evelyn described: "the joyfullest funeral that ever I saw for there was none that cried but dogs."

However, Cromwell did not rest in peace for long. In 1661, following the restoration of the monarchy, Cromwell was reviled as a regicide, and despised by the public for inflicting years of puritanical oppression. Cromwell and two other Parliamentary leaders, Henry Ireton and John Bradshaw, were exhumed, and their remains were dragged through the streets of London to Tyburn. Still in shrouds, they were hung from the gallows all day before being taken down and having the heads severed from their bodies—bodies that would have been in a state of considerable decay after several years in the grave. Cromwell's head was then fixed to a metal pole at Westminster Hall as a warning to anyone else who might dare oppose the monarchy. It remained on display there for more than twenty years.

In 1660, people who had participated in the trial and execution of Charles I were brought to trial. Ten people were condemned to death and publicly hanged, drawn, and quartered in London, and a further nineteen were sentenced to life imprisonment.

In the Museum of London, the death mask of Cromwell clearly shows a wart over the right eye.

Cromwell seems to have been quite fond of his warts and is reputed to have said the following to his portrait artist, Sir Peter Lely: "Mr. Lely, I desire you would use all your skill to paint my picture truly like me, and not flatter me at all; but remark all these roughnesses, pimples, warts, and everything as you see me, otherwise I will never pay a farthing for it."

can medicine be derived from corpses?

The dead have long been seen as potentially useful to the living, especially to help cure illness and prolong lives. As far back as 1550 B.C., we have records of human brains being cut up and used to cure eye ailments, according to an ancient Egyptian medical text.

In ancient Rome, epileptic patients were often prescribed several doses of human liver, which was usually taken from a gladiator, because gladiator livers were considered to be especially strong and courageous. The ancient Roman philosopher Celsus wrote in his book *De Medicina* that epileptics would be cured if they drank the blood of a slain gladiator.

Through the ages, in many different cultures, parts of dead bodies have been used as medicinal cures. For example, a cure for ulcers, goiters, cysts, and diseases such as tuberculosis was to touch a dead man's hand, preferably the hand of someone who had died a premature death. Public executions attracted large crowds, and people gathered in hope that a hangman would let them touch the hand of the still-warm victim as he dangled from the rope, and would pay the hangman for the privilege. The enterprising executioner at Newgate Prison in London developed a similar, and profitable, sideline—he chopped off his victim's hands and sold them. Hangmen also made money from selling remedies made of herbs and human fat from their victims.

During the 1600s and 1700s, people in Denmark believed that fresh blood would cure epilepsy. People flocked to public hangings and stood by the condemned, in the hope that they'd catch some blood in a cup. In the 1680s, human skulls were sold in London and people would grate them to a fine powder, which they would then consume. In 1721, the dispensatory of the Royal College of Physicans was recommending "three drams of human skull" to help epilepsy sufferers, and the *Pharmacopoeia Londinensis* of 1618—which was an encyclopedia of medicine and drugs—recommended powdered human skull for epilepsy. In German folk medicine, it was advised that women in labor should wear strips of tanned human skin as belts, to ease the pain of giving birth.

The incredibly well-preserved state of Egyptian mummies led European doctors from the Middle Ages onward to sell mummy parts as medication. In the 1100s, European doctors ground mummies into a powder, which they advised was to be drunk as a tea or used in a poultice, for its numerous health benefits. It was prescribed for many ailments, including coughs, nausea, epilepsy, migraines, incontinence, gout, bruises, and even fractures! When mummy was in short supply, unscrupulous mummy suppliers quickly mummified the bodies of executed prisoners to be sold as "ancient" mummies.

However, unsurprisingly, there is no evidence that eating any part of a mummy can actually help cure any ailment, so it's astonishing that its use continued until the seventeenth century. It fell out of favor mainly because of the problems associated with consuming it. French surgeon Ambrose Pare wrote that "not only does this wretched drug do no good, but it causes great pain to the stomach, gives foul-smelling breath, and brings on serious vomiting."

King Charles II (1630–85) used powder from Egyptian mummies and rubbed it onto his skin all over his body, as he believed it would promote longevity and transmit ancient greatness.

why did evita's corpse spend time in a wooden crate?

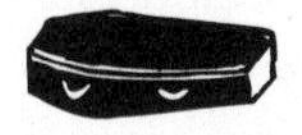

Eva Perón, who was popularly known as Evita, was the wife of the Argentine president Juan Perón. In 1952, at just thirty-three, she tragically died of cancer. Juan was so heartbroken about her death that he spent more than $73,700 (£46,000) having her embalmed. The man entrusted with this project was Dr. Pedro Ara, who spent a year embalming Evita, which involved endless injections of chemicals and placing the body into baths of acetate and potassium nitrate. He even applied layers of transparent plastic onto her face. When he had finished, it was said that she looked even better than she had in real life.

Perón had planned to build a mausoleum in Evita's honor, so that her body could be put on permanent public display. However, this plan was thwarted when his government was overthrown in 1955 by the army's General Aramburu, who led a successful coup against the government. Many army officers who initially professed loyalty to the government had lost in confidence in Perón and showed little enthusiasm for defending his regime. Perón was forced to flee Argentina, and lived in exile in Madrid, leaving Evita's body behind.

Because of Evita's enduring popularity, the new government didn't know what to do with the body, and so for security reasons they moved it several times. People protested and became extremely militant, demanding to view her enshrined body. Fearful of insurrection, General Aramburu arranged to smuggle her body out of the country and shipped it to Italy. For a while, the body was hidden in the apartment of an Argentine army major. Later, it was hidden in a wooden crate labeled "radio equipment" and stored in the attic of military intelligence headquarters. In 1957, the body was sent to a cemetery in Milan and was buried with the name Maria Maggi on the tomb.

Fueled by the mystery over the missing body, rumors began to circulate that Evita was still alive. In 1969, Aramburu was assassinated, and the continuing political unease eventually led to the restoration of Juan Perón to the presidency in 1973. Two years earlier, the location of Eva's body had been discovered, and the body had then been exhumed and smuggled to Madrid, to join Perón and his new wife, Isabelita. Dr. Ara was summoned once again, and the casket was opened with a crowbar. In his memoirs, Ara claimed that Evita's body was in good condition, except for some minor damage. However, in a statement released in 1985, Evita's two sisters contradicted Dr. Ara's account. They said that "gross mistreatment" of their sister's remains included hammer blows to the temple and forehead; an almost severed neck; a sunken nose and broken septum; cuts on the cheek, chest, and arm; an amputated finger; fractured kneecaps; and a layer of tar on the soles of her feet. Nonetheless, Dr. Ara restored the body, and in 1976, Evita was buried for the last time, in a vault in Buenos Aires, in the company of thirteen presidents of Argentina.

is it possible to turn dead bodies into diamonds?

A company based in the UK and United States called LifeGem is able to turn human ashes into diamonds. However, this service is by no means cheap—it can cost the customer about $19,200 (£12,000). The process takes about sixteen weeks and involves heating the ashes up to a temperature of 3,000 degrees centigrade, which converts the carbon in the ashes into graphite. Pressure is then applied to the graphite, turning it into a diamond. In fact, there is enough carbon in the average human body to make a necklace and a pair of earrings. LifeGem can also obtain carbon from a lock of hair, which means they can even create diamonds from people who prefer to have their bodies buried. The diamonds can be cut into various shapes and made into different colors, including blue, white, red, green, and yellow. They can even create diamonds from the ashes of your favorite kitty or pooch.

what gruesome event occurred while making an episode of *the six million dollar man*?

In 1977, while filming *The Six Million Dollar Man* at the Nu-Pike Amusement Park in Long Beach, California, one of the crew members was instructed to move a mannequin, which was painted fluorescent orange, dressed as a cowboy, and dangling from a rope. Unfortunately, as the crew member tried to lift the body down, one of its arms broke off, and the crew was horrified to discover that the "mannequin" was in fact a human corpse. The cowboy dummy turned out to be the embalmed body of Elmer McCurdy (1880–1911).

Amazingly, thanks to items found in McCurdy's mouth—a 1924 penny and a ticket from the Museum of Crime in Los Angeles—investigators were able to piece together his life story. McCurdy was a low-level criminal who bragged that he had killed a man in a brawl but whose only arrest was for being drunk in Kansas. After a variety of jobs, McCurdy decided to turn to robbing trains. In

1911, he traveled to Oklahoma and ambushed a train that he believed contained thousands of dollars of government tribal payments, but in fact it was just a passenger train, and he got away with just forty-six dollars and two jugs of whiskey. As law enforcers closed in, he warned them that he would not be taken alive. They quickly proved him right, as he was shot to death in the resulting shoot-out.

McCurdy's body was taken to a funeral parlor in Oklahoma. Nobody came forward to claim the body, so the undertaker came up with an ingenious idea—he charged customers a nickel to see "The Bandit Who Wouldn't Give Up." A while later, the undertaker was conned by two people who claimed to be relatives of McCurdy and wanted the body so that they could give him a proper and respectful send-off. The undertaker reluctantly handed over the body, which the conmen then used as an exhibit in a traveling carnival. Eventually, McCurdy's body ended up at the Nu-Pike Amusement Park, where it was stored and forgotten, until the film crew discovered the unusually lifelike mannequin. Once the mystery had been unraveled, McCurdy's body was returned to Oklahoma and buried in a cemetery in 1977. The state medical examiner ordered that two cubic yards of concrete be poured over McCurdy's coffin, so that his remains would never again be disturbed.

how did a nosebleed kill the "scourge of god"?

Attila the Hun (A.D. 403–A.D. 453) was renowned as one of the most fearsome warriors in history, earning himself the soubriquet the "Scourge of God." Ancient writers describe Attila as having a large head, a swarthy complexion, small, deep-seated eyes, a flat nose, a few hairs in the place of a beard, broad shoulders, and a short square body. Attila may have been no oil painting, but he had an impressive party trick—he could roll his eyes around and around in his head really fast. However, as befits a bloodthirsty warrior, his favorite hobby was hunting animals and then disemboweling them.

The Huns were a nomadic tribe that originated in Central Asia in the fourth century A.D., and settled in Eastern Europe. Legend has it that when a baby boy was born, to help him get accustomed to pain, a cut was made on his face. They were expert horse riders and could eat and sleep in their saddles. When they galloped into towns, they stole whatever they could and killed anyone in their way, including women and children.

The Hun empire was the strongest in Europe, stretching from Central Asia to as far west as Germany. After their uncle Ruga, the King of the Huns, died in 434, Attila and his brother Bleda ascended to the throne and became joint rulers of the Huns. However, Attila preferred to rule alone, so he arranged a hunting "accident" for his brother. Attila was now the sole leader of the Huns, and under his leadership the tribe terrorized the whole of Europe. When he conquered a city, Attila liked to display severed heads outside it.

On one of his wedding nights, Attila got very drunk—in fact, this happened a lot, as he reportedly had three hundred wives. However, on this occasion, he became so drunk that when he got a nosebleed after falling over, he choked to death on the blood. Rather than cry, the Huns mourned the loss of their leader by cutting off their hair and slashing their bodies with knives; they also cut their cheeks so they could cry tears of blood. After he was buried, the Huns executed the people who had buried him, so that his grave would never be discovered.

why was lenin embalmed?

Vladimir Ilyich Lenin (1870–1924) was the Russian leader who, along with Leon Trotsky, led the Bolshevik Revolution that toppled the Russian government in 1917. His leadership came to an end after he suffered a succession of strokes, which ultimately led to his death. Following his demise, Lenin's brain was removed, preserved, and then sliced up into thousands of sections. It was then analyzed by scientists, whose aim was to prove that he had an extraordinary mind. Some pieces even found their way to Germany during World War II, where the Nazis carried out their own tests on sections of Lenin's brain.

The rest of Lenin's body was put on display, and more than half a million people queued for hours in freezing temperatures to view his corpse. After witnessing the massive public response and the palpable grief, Lenin's successor, Joseph Stalin, realized the propaganda value of putting his dead predecessor's body on permanent display, which he believed would to keep the revolution on track.

For two months, doctors Boris Zbarsky and Vladimir Vorobiov worked night and day to preserve the body so that it looked as it had in life. They knew that failure would mean certain death, and so, perhaps unsurprisingly, they became world leaders in embalming methods. Mummification was achieved through soaking the body in baths of balsam, glycerine, and potassium acetate. Ever since, Lenin's remarkably well-preserved body has been housed in a mausoleum in Moscow's Red Square, and is checked regularly for deterioration. The body is dressed in a new suit every year, and has a yearly immersion in preserving chemicals. However, it has often been rumored that the body is actually made of wax.

After his death, Stalin was embalmed and displayed alongside Lenin. However, the brutality of Stalin's regime was repudiated by his successor, Nikita Khrushchev, and in 1961, Stalin's embalmed body was put into a coffin and buried by the Kremlin Wall.

where can you find a "bone chapel"?

If you fancy a gruesome family day out, why not visit the Chapel of Skulls in Czermna, Poland? As its name suggests, this church displays many bones, in fact, it is decorated with three thousand skulls and other bones, which line the chapel walls. The ceiling is densely covered with hundreds of human skulls, each positioned above two crossed leg bones. There are more than 20,000 more skulls stored in the vaults. The chapel was the innovative idea of a local Czech parish priest, who created it in 1776 to act as a memorial to those who had lost their lives in the wars of the 1600s and 1700s. This popular attraction also contains the bones of some of the victims of the many cholera epidemics that plagued the area.

Another macabre chapel, called the "Capela dos Ossos" or "Bone Chapel," can be found in Evora, Portugal. It was built in the fifteenth century by Franciscan monks, and measures 61 feet (18.7 m) long and 36 feet (11 m) wide, with three small windows and eight pillars. The chapel is decorated with human bones and skulls from about five thousand bodies taken from local cemeteries. The bones are held together by cement and can be found covering the walls and pillars of the chapel. The chapel also contains two-dried out leathery corpses hanging from the ceiling, but their identities are unknown.

Beneath the city of Rome lie hundreds of tunnels dug out of the soft volcanic rock, which extend for miles and contain the buried remains of tens of thousands of people. Rows of tightly packed bones are piled from the floor to the ceiling, and have remained there for almost two thousand years. In the second century, Christians started burying their dead underground, and the catacombs continued to function as cemeteries until the fifth century. The bones mostly belong to Christians, but the catacombs also house the bones of Jews and pagans. Burials of early Christians involved wrapping the body in a sheet or shroud and placing it into a niche cut out of the rock. The niche would then be sealed with tiles and fixed with mortar or a slab of marble, upon which the name of the deceased would sometimes be engraved. Some of the tunnels are open to the public for guided tours.

what is "promession"?

Cremation was introduced in response to the growing shortage of available land for burial, with the first legal cremation being carried out in the Cremation Society's crematorium at Woking, Surrey, in 1885. The cremation process has progressed from the first coke-fired ovens to more modern gas and electric cremators over the last one hundred years. Nowadays, almost all cremators use gas, but this unfortunately pollutes the environment. Ovens in the crematoria burn for about an hour and half per body, at temperatures of up to 1100°C, using up the same amount of energy as a typical home would in a month. The burning of wooden coffins also causes harmful emissions to be released into the atmosphere. Another problem is that the dental fillings found in the teeth of the cremated can emit mercury and contaminate the environment. However, crematoria are now being fitted with special filters to eliminate this problem. Still, if you're hoping for an eco-friendly death, there is a greener way to dispose of your earthly remains.

An enterprising Swedish woman, environmental

biologist Susanne Wiigh-Masak, has come up with an eco-friendly alternative to cremation, with a method known as "promession." Promession involves submerging the corpse in liquid nitrogen, which causes the body to become brittle. It is then vibrated in a machine, where it shatters into millions of pieces. While the machine vibrates, a metal separator is used to find items such as metal plates, dental fillings, and hip-replacement parts. Eventually, all that is left is a powdery substance, which is then placed into a box made of corn or potato starch, which can then be buried in a shallow grave. The box and its contents will then turn into compost over a period of about six to twelve months. Often, relatives choose to plant a tree or shrub at the grave site, to make a pleasingly green use of the compost.

what was alistair cooke's last big story?

Alistair Cooke (1908–2004) was a renowned journalist and broadcaster who was perhaps best known for *Letter from America*, a weekly fifteen-minute radio program that was broadcast on BBC Radio Four and the World Service from 1946 to 2004, making it the world's longest-running speech radio program. Each week, Cooke dealt with a topical American issue, often in an anecdotal and humorous fashion. Cooke was educated at Jesus College, Cambridge, but later in life he became a naturalized American citizen and lived in a rent-controlled Manhattan apartment for more than fifty years. In March 2004, Cooke died at his home in New York. He was cremated, and his ashes were scattered in Central Park.

In December 2005, the New York *Daily News* reported that Cooke's bones had been removed from his corpse prior to cremation, by the staff of Biomedical Tissue Services (BTS), a "tissue-recovery" company based in Fort Lee, New Jersey. The report revealed that the company was being investigated for a range of gruesome practices involving the illegal harvesting of human bones and tissues. The company had been taking body parts from funeral homes, often paying the homes $1,000 (£626) or more per corpse, and selling the parts to transplant companies around the world for dental implants, hip replacements, and other procedures. BTS would forge family consent and other donor forms, often going against the family's expressed wishes. They also engaged in a number of unusual and unsafe practices, including failing to screen body parts for diseases such as HIV/AIDS. In one case, the company replaced the bones of a woman from Queens,

New York, with PVC piping, to trick members of her family.

Cooke had died at the age of ninety-five from lung cancer, which had spread to his bones, but this did not prevent BTS from selling his bones for use as bone grafts, even though the cancer meant they were unsuitable for this type of therapy. The company had harvested his bones the day after he died, and sold them for $11,000 (£6,900). His remains were given to the family for cremation. He was considered a suitable donor only because BTS had lied about the cause of death, listing it as a heart attack rather than cancer and claiming that Cooke was eighty-five when he died, not ninety-five. Fortunately, thanks to the police investigation, Cooke's bones were never implanted in anyone.

In February 2006, Dr. Michael Mastromarino, the head of BTS, was convicted, along with three employees, of stealing body parts from about one thousand four hundred corpses. Mastomarino was sentenced in June 2008 to between eighteen and fifty-four years in prison.

which hollywood star was kidnapped after his death?

Hollywood legend and star of silent films Charlie Chaplin died in 1977 and was laid to rest in Switzerland. However, three months later, his widow, Oona, received a strange phone call from a man demanding $600,000 (£375,000) for Charlie's body to be returned to her. She made inquiries, and indeed it was true, Charlie's body had been stolen. One night, two people had broken into the cemetery at Corsier-sur-Vevey, dug up his body, and carried it away.

Oona contacted the police, and they tapped her phone line. She continued to maintain contact with the body snatchers but refused to pay the ransom. After several more calls, the criminals reduced the ransom down to $250,000 (£156,000), but by this point the police had traced them to a phone in Lausanne. The kidnapper turned out to be a twenty-four-year-old Polish car mechanic and, together with a friend, admitted to stealing the body. The two men were sent to prison, and Charlie's body was reburied in a steel vault, under 6 feet (2 m) of concrete, to ensure that it would not happen again.

why was the tomb of st. thomas becket at canterbury one of england's most popular pilgrimage sites?

Thomas Becket was born circa 1118 at Cheapside, London, to the prosperous London merchant Gilbert Becket. Through his father's wealthy connections, Becket received an excellent education in London, Paris, Bologna, and Auxerre. On his return to England, he became an agent for the Archbishop of Canterbury, Theobald, who sent him on several important missions to Rome. Becket was clearly an impressive young man, as Theobald recommended him to the King, Henry II. Henry made Becket his Lord Chancellor, and the two became close friends. Henry was determined to diminish the power of the Church, and used Becket to enforce land taxes against the clergy. After the death of Theobald in 1162, Henry appointed Becket to the post of Archbishop of Canterbury.

Until this point, Becket had been a fun-loving courtier, but after his appointment he became an ascetic, which meant he pursued a simple, serious-minded life devoted to his faith, abstaining from worldly pleasures. A rift developed between Becket and Henry, as it became clear that Becket would not be the puppet Archbishop Henry had hoped for but instead would defend the Church's interests and wealth. At the assemblies at Clarendon Palace in 1164, Henry demanded that Becket sign documents recognizing the King's authority over the Church, but Becket refused. This meant the two power bases were now effectively at war, and Becket was forced to flee the continent, where he remained in exile for several years. Nonetheless, he continued to cause trouble for Henry, and when, in 1170, Becket excommunicated three leading bishops, Henry was reported to have lifted his head from his sickbed and moaned, "Will no one rid me of this troublesome priest?" Four of Henry's knights interpreted this as a royal command and brutally murdered Becket at Canterbury Cathedral, in a crime that shocked Christian Europe. Becket soon became regarded across Europe as a religious martyr, and just three years after his death, he was canonized by Pope Alexander III.

In the years that followed, Becket's tomb at St. Dunstan's, Canterbury, became one of the most popular pilgrimage sites in England. Holy tourism was a lucrative business for the Church, and Becket's tomb was particularly appealing for a number of reasons. First, it was much easier to get to than Rome or Santiago de Compostela, being a reasonably short journey from London; and because of Becket's status, the "indulgences," or religious benefits, were just as great. Second, the pilgrimage route was an interesting and relevant one, as it traced Becket's final journey, from St. Mary's Priory in Southwark, where he gave his last sermon, past landmarks such as "Becket's Well" in Otford, Kent, where Becket was said to have created two natural springs by striking the ground with his crosier. The third reason for the tomb's enormous appeal was surely that it was extremely and fascinatingly gruesome. Becket's death had been a violent one, and visitors to Canterbury could actually get to see where the sword had split his skull in two!

how much coffee would it take to kill a man?

According to legend, coffee was discovered by a goatherd named Kaldi in ninth century Abyssinia. Kaldi is said to have noticed that his goats became unusually lively after eating the berries from a particular shrub, so he decided to try the berries himself, and found that he too became more alert and energetic. This was of course because the berries were coffee beans, which contain caffeine, which is a stimulant. Caffeine is the world's most widely used psychoactive substance.

In moderate doses, caffeine is harmless and can even have beneficial effects. Studies have shown that cyclists and runners who have consumed caffeine can perform at a competitive pace for significantly longer periods and distances. However, excessive use of caffeine over time can lead to a condition called "caffeinism," whose symptoms can include muscle twitching, headaches, hyperventilation, insomnia, and heart palpitations.

It is also possible to overdose on caffeine, which can lead to a wide range of symptoms, including flushing, nervousness, increased urination, delirium, muscle twitching, and an irregular heartbeat. Extreme cases of caffeine overdose can even cause death, but this would be very unlikely to happen through drinking coffee alone. To cause a fatal overdose, a person would need to consume more than 0.01 pounds (4.5 g) of caffeine, which would mean drinking about eighty to one hundred cups of coffee within a period of about four to five hours. There have been reported cases of deaths from caffeine overdose, but these were believed to have been caused by caffeine pills.

can a person die of a broken heart?

James Callaghan and Audrey Moulton met while they were in their teens, at the Baptist church where they both taught Sunday school. They married in 1938, and Callaghan went on to become the only man to hold all four of the "Great Offices of State": prime minister, chancellor of the exchecker, home secretary, and foreign secretary. When, in her later years, Audrey developed Alzheimer's disease, Callaghan devoted himself to her care, saying, "For sixty years she gave up everything for me, and now I'm going to give up everything for her." She died on March 15, 2005, and Callaghan followed just eleven days later.

Could Callaghan have died of a broken heart? In 1969, a study was published in the *British Medical Journal.* Scientists had monitored four thousand five hundred widows for nine years after their husbands died, and found that the widows were 40 percent more likely to die in the six months after a spouse's death. After that period, the risk slowly fell back to normal levels. Most of the widows had died of a heart attack. Numerous studies have since confirmed that people are more at risk of death in the months following the loss of a spouse, and that there is an increased risk of heart disorders.

In 2005, researchers at Johns Hopkins University in Baltimore found that sudden emotional stress could cause a condition called acute stress cardiomyopathy, or "broken-heart syndrome." The researchers examined nineteen patients, mostly women in their sixties and seventies, who were all displaying symptoms of conventional heart attacks but with a number of key differences. Normally, heart attacks are caused by damage to the heart muscle, but these patients showed no heart tissue damage. None of them had a history of heart problems or chronic stress, and all of them had just received some kind of emotional trauma or sudden shock, including a death in the family, an armed robbery, and, in one case, a surprise birthday party. Each of the patients had unusually high levels of stress hormones in their blood, and the researchers determined that it was these hormones, including adrenaline, that were impairing the heart's ability to pump.

> *Ah, what a trifle is a heart,*
> *If once into love's hands it come!*
>
> JOHN DONNE, *"The Broken Heart"*

why do insects fall prey to pitcher plants?

Around the world, there are more than six hundred different species of carnivorous plants. One group of these are called "pitcher plants," of which the two main families are the Nepenthaceae and

Sarraceniaceae. Pitcher plants are commonly found in the United States, and come in an array of colors, including yellow, green, red, and white, and some can grow as high as 4 feet (1 m) tall.

All pitcher plants work by catching their prey in an unusually shaped, hollow leaf structure, which looks like a pitcher or a jug. First, the plant has to attract the insects. For this purpose, pitcher plants have a lid or hood over the pitcher, which contains nectar. This lid also serves to prevent the pitcher from filling with rainwater. Once the insect is inside, hairs inside the lid direct the insect deeper into the plant. As it explores further, the insect will suddenly reach a smooth, waxy area, found on the upper part of the pitcher, which causes it to slide into the liquid below and drown. The insect is unable to climb out of the pitcher, because the hairs inside it grow downward and act as spears, blocking its path. The liquid the insect falls into contains acids and enzymes to help it digest food. When the insect falls into this liquid, its soft body parts will slowly dissolve in these digestive juices. Fungi and bacteria, which also help to break down the insect's bodies, are also found inside pitcher plants.

The "yellow pitcher plant" (*Sarracenia flava*), also known as the "trumpet pitcher plant" because of its trumpet-shaped leaves, ensures its insect victims cannot escape by lacing its nectar with "coniine," a drug that is narcotic to insects, and paralyzes them. The insect then falls into the trumpet pitcher's liquid and is digested. As rotting insect bodies fill the plant, the smell attracts other potential victims: blowflies. This plant is so effective at catching prey that it may contain as many as a thousand insects; so many, in fact, that the plant can become too heavy and topples over. Sometimes even frogs have become prey after slipping into the plant while hunting insects.

However, there are some insects that do not fall victim to the pitcher plant but instead use it for food. Some spiders produce webs across the pitcher's leaf opening, so that when insects fly in to drink nectar, they become caught in the web, providing the spider with a meal. Some mosquitoes and flies lay their eggs in the pitcher's liquid. These eggs are unaffected by the plant's digestive juices, and when they hatch, the larvae find themselves surrounded by rotting insect parts on which to feast.

Although the main diet of pitcher plants is typically bugs and insects, there is one variety that sometimes preys on even larger creatures. The "tropical pitcher plant" (*Nepenthes*), which is found mainly in the rainforests of Southeast Asia, contains some pitchers as large as footballs and has been known to devour whole birds, frogs, and even rats.

what happened when william the conqueror's corpse was placed into his tomb?

William the Conqueror (ca. 1028–87) became King William I following his conquest of England at the Battle of Hastings. As a young adult, William was so fit that he could jump onto his horse when dressed in full armor. However, later in life he piled on the pounds and became grossly overweight. When King Philip of France made some spiteful comments about William being too fat, William set fire to the small French town of Mantes in France. As the fire burned, his horse stepped on a hot cinder and slipped. This caused William to be thrown forward and fall onto the point of this saddle, therefore sustaining a nasty injury that caused internal bleeding. He died six weeks later.

As William was large and decomposition of his body caused it to swell, it became too big to fit inside the stone tomb. Legend has it that two soldiers had to stand on the body to push it in. In their desperation to cram it in, they jumped up and down on the body and broke its spine. The broken spine pierced the stomach, and because of the accumulation of gases, it exploded, splattering the poor soldiers with rotting flesh. The smell was so putrid that everyone, including priests, rushed out of the church to get some fresh air.

about the author

Francesca Gould is the author of *Why You Shouldn't Eat Your Boogers and Other Useless or Gross Information About Your Body* and several successful British books on the subject of anatomy and holistic health. She lives in Bristol, England.

Another must-read for those in search of useless or gross facts!

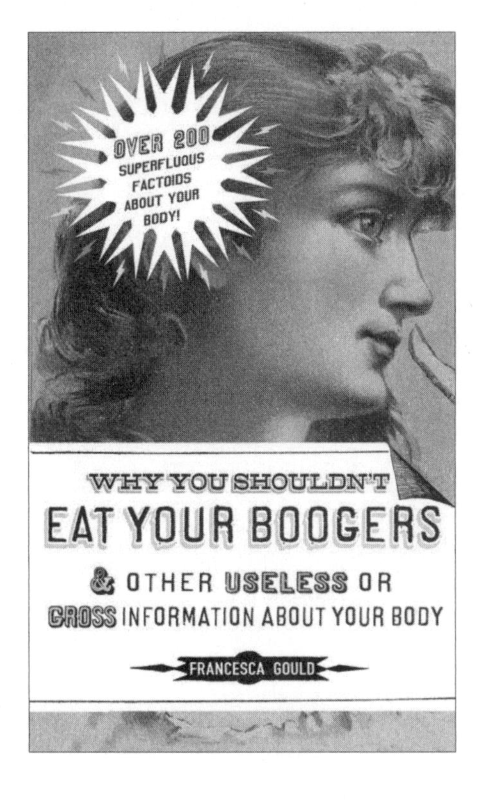

A delightful, sometimes embarrassing, yet always hilarious catalog of useless or gross pieces of information about the human body

ISBN 978-1-58542-645-4